Cultural Diversity in the Armed Forces

This book addresses the different ways in which armed forces around the world respond to the cultural diversity of their parent societies, in relation to ethnicity and gender in particular, in order to enhance their quality and legitimacy.

While ethnicity and gender are the leading areas of diversity examined here, several others are also addressed, including sexual orientation, language, religion, region and class. Analysing the situation in fourteen countries, ranging form North and South America via Africa, India and Israel, to Western Europe, the book finds that the tension between military necessity and organizational effectiveness, on the one hand, and political and social desiderata, on the other, is present in all the cases under consideration. While this volume demonstrates that an even-handed representation of ethnic minorities and gender is now widely recognized as crucial for enhancing the armed forces' social legitimacy and professional quality, it also analyses how and why a number of issues remain.

This book ultimately supports the argument for a diverse military force without denying that it has always been a complicated, if not paradoxical, ideal in practice.

The book will be of interest to students of minority studies, military sociology, gender and political science.

Joseph Soeters is a professor in organization studies at the Netherlands Defence Academy and Tilburg University. **Jan van der Meulen** is an associate professor at the Netherlands Defence Academy.

Cass military studies

Intelligence Activities in Ancient Rome
Trust in the gods, but verify
Rose Mary Sheldon

Clausewitz and African War
Politics and strategy in Liberia and Somalia
Isabelle Duyvesteyn

Strategy and Politics in the Middle East, 1954–60
Defending the northern tier
Michael Cohen

The Cuban Intervention in Angola, 1965–1991
From Che Guevara to Cuito Cuanavale
Edward George

Military Leadership in the British Civil Wars, 1642–1651
'The genius of this age'
Stanley Carpenter

Israel's Reprisal Policy, 1953–1956
The dynamics of military retaliation
Ze'ev Drory

Bosnia and Herzegovina in the Second World War
Enver Redzic

Leaders in War
West Point remembers the 1991 Gulf War
Edited by Frederick Kagan and Christian Kubik

Khedive Ismail's Army
John Dunn

Yugoslav Military Industry, 1918–1991
Amadeo Watkins

Corporal Hitler and the Great War, 1914–1918
The List Regiment
John Williams

Rostóv in the Russian Civil War, 1917–1920
The key to victory
Brian Murphy

The Tet Effect, Intelligence and the Public Perception of War
Jake Blood

The US Military Profession into the 21st Century
War, peace and politics
Edited by Sam C. Sarkesian and Robert E. Connor, Jr

Civil–Military Relations in Europe
Learning from crisis and institutional change
Edited by Hans Born, Marina Caparini, Karl Haltiner and Jürgen Kuhlmann

Strategic Culture and Ways of War
Lawrence Sondhaus

Military Unionism in the Post-Cold War Era
A future reality?
Edited by Richard Bartle and Lindy Heinecken

Warriors and Politicians
U.S. civil–military relations under stress
Charles A. Stevenson

Military Honour and the Conduct of War
From ancient Greece to Iraq
Paul Robinson

Military Industry and Regional Defense Policy
India, Iraq and Israel
Timothy D. Hoyt

Managing Defence in a Democracy
Edited by Laura R. Cleary and Teri McConville

Gender and the Military
Women in the armed forces of Western democracies
Helena Carreiras

Social Sciences and the Military
An interdisciplinary overview
Edited by Giuseppe Caforio

Cultural Diversity in the Armed Forces
An international comparison
Edited by Joseph Soeters and Jan van der Meulen

Cultural Diversity in the Armed Forces
An international comparison

**Edited by Joseph Soeters and
Jan van der Meulen**

LONDON AND NEW YORK

First published 2007
by Routledge
2 Park Square, Milton Park, Abingdon, Oxon OX14 4RN

Simultaneously published in the USA and Canada
by Routledge
270 Madison Ave, New York, NY 10016

Routledge is an imprint of the Taylor & Francis Group, an informa business

Transferred to Digital Printing 2009

© 2007 Joseph Soeters and Jan van der Meulen for selection and editorial matter; individual chapters, the contributors

Typeset in Times by Wearset Ltd, Boldon, Tyne and Wear

All rights reserved. No part of this book may be reprinted or reproduced or utilized in any form or by any electronic, mechanical, or other means, now known or hereafter invented, including photocopying and recording, or in any information storage or retrieval system, without permission in writing from the publishers.

British Library Cataloguing in Publication Data
A catalogue record for this book is available from the British Library

Library of Congress Cataloging in Publication Data
A catalog record for this book has been requested

ISBN10: 0-415-39202-0 (hbk)
ISBN10: 0-415-54510-2 (pbk)
ISBN10: 0-203-96740-2 (ebk)

ISBN13: 978-0-415-39202-0 (hbk)
ISBN13: 978-0-415-54510-5 (pbk)
ISBN13: 978-0-203-96740-9 (ebk)

Contents

List of tables	ix
List of contributors	xi
Preface	xiv

1 Introduction   1
JAN VAN DER MEULEN AND JOSEPH SOETERS

2 Diversity in the armed forces of the United States   15
CHARLES MOSKOS

3 Diversity in the Canadian forces   31
DONNA WINSLOW, PHYLLIS BROWNE AND
ANGELA FEBBRARO

**4 Indigenous integration into the Bolivian and
Ecuadorean armed forces**   48
BRIAN R. SELMESKI

5 Diversity in the Brazilian armed forces   64
CELSO CASTRO

6 Diversity in the South African armed forces   77
LINDY HEINECKEN

7 Diversity in the Eritrean armed forces   95
MUSSIE TECLEMICHAEL TESSEMA

8 Diversity in the Indian armed forces   111
LEENA PARMAR

9 **Diversity in the Israel Defense Forces**   125
   EDNA LOMSKY-FEDER AND EYAL BEN-ARI

10 **Ethnic diversity in the British armed forces**   140
   CHRISTOPHER DANDEKER AND DAVID MASON

11 **Diversity in the French armed forces**   154
   BERNARD BOËNE AND CLAUDE WEBER

12 **Diversity in the German armed forces**   171
   HEIKO BIEHL, PAUL KLEIN AND GERHARD KÜMMEL

13 **Diversity in the Belgian armed forces**   185
   PHILIPPE MANIGART

14 **Diversity in the Dutch armed forces**   200
   RUDY RICHARDSON, JOLANDA BOSCH AND
   RENÉ MOELKER

   Index   215

Tables

2.1	Blacks as a percentage of US armed forces, selected years 1949–2004	17
2.2	Blacks as a percentage of total personnel by grade and service, 2004	18
2.3	Females as a percentage of US armed forces, selected years 1970–2004	21
2.4	Females as a percentage of total personnel by grade and service, 2004	22
2.5	Percentage married, by rank and gender, 2004	25
2.6	Percentage of female service members by race, 1980, 1998 and 2004	26
2.7	Race/gender contrasts in the military	27
3.1	2001 Census data and demographic trends for designated groups in the Canadian armed forces	35
5.1	Regional background of Brazilian officer-cadets	69
5.2	Religious background of Brazilian officer-cadets	70
5.3	Social background of Brazilian officer-cadets	71
5.4	Cadets' parents' background	72
5.5	Cadets' parents' background: upper versus lower military	73
6.1	Composition of the South African National Defence Force by former force	81
6.2	Composition of the South African National Defence Force by race and gender	86
7.1	Ethnic profile of the Eritrean People's Liberation Front's army, 1993	97
7.2	Ethnic profile of the Eritrean armed forces, 2004	102
7.3	Representation of women in the Eritrean armed forces (%)	105
10.1	Intake to untrained strength of UK Regular forces by ethnic origin and service (%)	143
12.1	Number of resettlers immigrating to Germany, 1981–2004	173
12.2	Number of female soldiers in the three German services according to career, 2004	181

x *Tables*

13.1	Personnel distribution by language in the Belgian armed forces, 2005 (%)	187
13.2	Distribution of foreigners by rank and nationality, 2005	189
13.3	Distribution of women among categories in the Belgian armed forces, 1976–2005	191
13.4	Occupational distribution of female officers, by force, 2005	192
14.1	Division of ethnic minorities in the Netherlands	202
14.2	Ethnic minority groups within the Dutch armed forces (%)	203
14.3	Organizational presence of female military personnel (%)	207
14.4	Percentage of women by occupation in the Dutch armed forces, 2004	209

Contributors

Eyal Ben-Ari is professor of anthropology at the Hebrew University of Jerusalem. He has conducted research on the Israel Defence Forces and the contemporary Japanese military and peacekeeping forces. In addition, he has recently carried out fieldwork in a Japanese kindergarten.

Heiko Biehl is a political scientist and a senior researcher at the Bundeswehr Academy for Information and Communication in Strausberg, Germany. Among other topics he has published about morale and cohesion in military units.

Bernard Boëne is professor of sociology at the University of Rennes and director of social science research at the French military academy, Saint-Cyr Coëtquidan, where he taught for over two decades. His publications all centre on military action, martial institutions, and their relations with state and society in a comparative perspective.

Jolanda Bosch is a psychologist at the Faculty of Military Sciences of the Netherlands Defence Academy. Her research and teaching focus on gender issues. She has been the chair of the Women's Defence Network and she regularly advises on evolving policies with regard to diversity.

Phyllis Browne is a graduate of McGill and Concordia Universities in Montreal, Canada. Her doctoral degree is in sociology and her main areas of interest are education, gender issues, labour markets and cultural diversity. She is currently employed by the federal government of Canada.

Celso Castro is a professor and current director of the Centro de Pesquisa e Documentação de História Contemporânea do Brazil (Centre for Documentation and Research in Brazilian Contemporary History) of Fundação Getulio Vargas, in Rio de Janeiro. He has published many books and articles about the military in Brazilian history and society.

Christopher Dandeker is professor of military sociology and head of the School of Social Science and Public Policy at King's College London and co-Director of the King's Centre for Military Health Research. He has published books and articles in the field of civil–military relations.

Angela Febbraro received her Ph.D. in applied social psychology at the University of Guelph, Ontario, Canada, in 1998. She is a defence scientist at Defence R&D in Toronto. Her current research interests focus primarily on diversity issues in the military.

Lindy Heinecken was formerly affiliated with the South African Military Academy. Currently she is a researcher and teacher at the University of Stellenbosch. Her main research interests focus on civil–military relations in general and diversity issues in particular.

Paul Klein studied psychology and sociology and until recently was employed as a deputy director and senior researcher by the Bundeswehr Institute of Social Research in Strausberg, Germany. He is also a lecturer at the University of the Bundeswehr in Munich.

Gerhard Kümmel is a senior researcher at the Bundeswehr Institute of Social Research in Strausberg, Germany. He has published extensively on a range of civil–military topics. He is the executive secretary of the research committee on armed forces and society of the International Sociological Association.

Edna Lomsky-Feder is a senior lecturer at the School of Education of the Hebrew University of Jerusalem. She has conducted research on war and the military in Israel, and published extensively on this topic.

Philippe Manigart is professor of sociology and head of the Department of Behavioural Sciences at Brussels Royal Military Academy. He has an MA in sociology from the University of Chicago and a Ph.D. from the Free University of Brussels.

David Mason is professor of sociology and dean of the School of Social Sciences at Nottingham Trent University, UK. He has published extensively on labour market and diversity issues. His recent work includes research into the recruitment of the minority nurses and on diversity in the armed forces.

Jan van der Meulen is a sociologist who works as an associate professor at the Faculty of Military Sciences of the Netherlands Defence Academy. His publications focus on public opinion about armed forces and military missions.

Rene Moelker is an associate professor of sociology at the Faculty of Military Sciences of the Netherlands Defence Academy. His work concentrates on military families, military technology, the military profession and the military sociology of Norbert Elias.

Charles Moskos is professor of sociology at Northwestern University in Evanston, Illinois. His research on military personnel has taken him to Vietnam, Cyprus, Somalia, Haiti, Bosnia, Kosovo and Iraq. In 1999 he was elected to the American Academy of Arts and Sciences.

Leena Parmar is a sociologist and associate professor who is employed by

Rajasthan University. She is head of the Department of Sociology. She has published on a range of civil–military themes with regard to the Indian armed forces.

Rudy Richardson is assistant professor at the Faculty of Military Sciences of the Netherlands Defence Academy. His work concentrates on managing cultural diversity, human factors and safety, military culture, policy development and research methodology. He is also chairman of the Dutch Defence Multicultural Network.

Brian R. Selmeski is a Research Associate in Cultural Anthropology at the Royal Military College of Canada's Centre for Security, Armed Forces, and Society. His research areas include military diversity, institutional culture and Latin America.

Joseph Soeters is professor in organization studies at the Netherlands Defence Academy and Tilburg University. His research focuses on international military cooperation and national styles of operation, including command and control, communication with local populations, and various organizational issues.

Mussie Teclemichael Tessema completed his undergraduate studies in business management at the University of Asmara, Eritrea (1995). He graduated from the Institute of Social Studies, The Hague, the Netherlands (1998), and then worked as a lecturer at the University of Asmara. He received his Ph.D. at the University of Tilburg for a study of public organizations.

Claude Weber is Maître de Conférences (professor and researcher) in sociology and director of the sociology department at the French military academy, Saint-Cyr Coëtquidan. He teaches military sociology, intercultural relations, management in professional armies, sociology of military action and military public communication.

Donna Winslow until recently held the chair of social anthropology (development and social transformation processes) at the Vrije Universiteit in Amsterdam. She received her Ph.D. from Université de Montréal. She has conducted field research in theatre with Canadian units in the former Yugoslavia, the Golan Heights and Afghanistan.

Preface

In 1999 the edited volume *Managing Diversity in the Armed Forces* was published by Tilburg University Press. Notwithstanding its limited distribution, it received quite some attention. Colleagues from all over the world responded with kind and stimulating comments. To us, these reactions were an important incentive to edit an updated, revised and expanded new edition. However, the most important reason for doing that lies in the continuing relevance of the subject itself. In fact, since 1999 cultural diversity as a worldwide topic has only become more prominent, not without dramatic overtones. While armed forces do not necessarily and immediately reflect all the tensions and moods of society, they are hardly immune to them. Sooner or later the military will experience their own share of the differences and inequalities within their parent societies and will have to work out solutions befitting democratic civil–military relations. It is our hope that this book contributes to the finding and implementing of those solutions.

We are grateful that Routledge has been willing to publish this book. We especially want to thank the Netherlands Defence Academy for granting us the time necessary for doing the numerous things editors run into. Most of all, we thank the authors, either for updating their chapters or for writing new ones. While every author is responsible for the substance of his or her own chapter, we take full responsibility for the concept and the quality of this volume as a whole.

Jan van der Meulen
Joseph Soeters

1 Introduction

Jan van der Meulen and Joseph Soeters

Introduction

Since the beginning of 2006, policemen and policewomen in Western Australia have won the right, while on duty, to wear a blue turban or a blue scarf, with an official badge attached to it. This new measure presents Sikhs and Muslims with an alternative to the standard police cap. The government of Western Australia argued that the police organization should mirror society, in order to enhance ethnic communities' feeling of being represented. The police union, however, voiced the fear that citizens might resent being confronted with constables wearing 'exotic' uniforms.[1]

In earlier times a newspaper report like this might have sounded innocent and slightly colourful. Not any more: cultural diversity has become a major theme in the contemporary world, emotionally charged and politically loaded. Whether in terms of ethnicity, religion or gender – in the above example all three of them come into play – core social identities are at stake, touching as well as contesting the basics of collective and individual self-defining. In principle, any message from anywhere on this topic can be recognized and will be unpacked in the context of local circumstances. As a result, we are witnessing, in the words of Berger and Huntington, a pattern of 'many globalizations'.[2]

Cultural diversity is not just another topic. Meanwhile, the military are not just another organization – at least, not all of the military, all of the time. The management of violence and threat, the application of force and ultimately the conduct of war are still defining the military's core business, and, as a consequence, the mainstream of military culture. Whether and how organizational, social and political developments in society at large fit in with this typical, presumably unique, culture can be looked upon as one of the classical themes of military social science. Accommodating armed forces to diversity very much belongs to that theme and has, in fact, gained quite some prominence as a battleground for conflicting views on the best civil–military fit. From the end of the 1990s onwards the optimistic metaphor of a diverse 'rainbow' military, reflecting a multicultural 'rainbow' society, has been juxtaposed with a rather more gloomy vision of 'culture wars' being imported into the armed forces.[3]

In this volume, ethnicity is one of the two leading types of diversity to be discussed in relation to armed forces from a range of countries around the world, gender being the other. In this chapter we will start by briefly introducing the historical status of both cases within the context of civil–military relations. Thereafter we will argue in some detail why it is proving ever more important for armed forces to aim to recruit and succesfully manage a diverse organization. Clearly, present-day circumstances bring with them difficulties in trying to do so. Multicultural strains are having consequences for society and armed forces. These may be confronting and painful, especially for the members of minority groups themselves.

We do not want to eschew these kinds of uncomfortable experiences and we will discuss some examples. Doing so will give us an opportunity to point out a number of insights and tools that the literature on diversity management offers for dealing with strains of this kind in the workforce. It is our conviction that the application of these insights is quite relevant for the military and for fostering unit effectiveness.

In the last section of this introduction we will give a brief overview of the country chapters that constitute the substance of this volume. We will explain the spectrum of diversities to be addressed in addition to ethnicity and gender.

Ethnic soldiers

While everyone in the United States was arguing about whether a black man could fly an aircraft, in 1941 the First Lady, Eleanor Roosevelt, climbed into the cockpit with a black pilot, C. Alfred 'Chief' Anderson, for an aerial look at Alabama. Even after this vote of confidence, it still took another year and a lawsuit before the War Department finally announced the formation of an all-black flying unit. An unwelcome sight to many, these flying personnel broke yet another race barrier and dispelled the myth that 'blacks can't fly'.[4] The story of the so-called Tuskegee Airmen from Alabama – highly successful black pilots in the Second World War – is a historic case about diversity in the armed forces. It testifies to the controversies that surround diversity, the difficulties it entails and the successful ways in which it may be managed in the long run.

History offers many more examples of the ways in which multi-ethnic societies have translated and controlled their diversity in the context of armed forces. The demographic composition of the Ottoman armed forces, for example, continued to give rise to heated debates for decades, and even centuries, about whether or not non-Muslim citizens should be allowed (or required) to enlist as soldiers and to go to the battlefield.[5]

In her seminal book *Ethnic Soldiers*, Enloe depicted an impressive array of cases drawn from around the world. From India to the Philippines, from Canada to Iraq, from the United States to the Soviet Union, from Yugoslavia to Lebanon, Enloe presented a wealth of material. Her multilevel analyses, touching on the use of 'martial races', on conscription and the rhetorics of nation-building, on the entwinement of security policies and ethnic organization, and on many more related themes, provide a fascinating read, still very relevant for today's purposes.[6]

Other books have explored the same themes while adding still more cases, from the multinational Habsburg Army to the French colonial Army[7] and from South Africa via Singapore to Israel.[8] In these studies the political implications of ethnic diversity in the military weigh heavy. The 'Trojan horse dilemma'[9] lies at the heart of much of the topical literature. That particular theme is not absent in the present volume, though with clear differences from one country to another. Overall, the emphasis in this book is rather more 'cultural' than 'political', granted that the two angles are not always easy to separate. As we have already hinted, doing that is becoming more and more difficult.

Women warriors[10]

The history of women in the military is quite different from that of ethnic minority soldiers and their past. While the latter were abundantly present and often in demand, albeit conditionally, until recently women have been excluded from armed forces. A few notable if not mythological exceptions notwithstanding, it was only in the course of the twentieth century that women started to wear military uniforms. And it was only some twenty-five years ago, on average, that separate female corps began to lose ground to more or less full integration.

Today, in more and more countries women have a formal right to enter all, or almost all, ranks and positions in the military, sometimes including special forces and submarine crews. As more and more positions are opened up, gradually the number of female soldiers is increasing. In quite a number of countries female generals are no longer exceptional. Undoubtedly these developments are having a profound impact on armed forces and on civil–military relations.

This is not to say that full gender equality in numbers and opportunities exists in the armed forces across the world; far from it. Nor can it be argued that the interaction between military men and women on vessels and in bases and garrisons is always developing smoothly. There are simply too many reports of incidents of sexual harassment ocurring all over the world. On a more fundamental level, to varying degrees male identities are still constructed around patriarchism and machismo.[11] The notion of an ingrained incompatibility of womanhood and soldierhood is far from having disappeared.

As in the case of ethnic diversity, general developments concerning gender are relevant while one is talking about the military. While the backlash of ethnic tensions looks worrisome, the speed of female emancipation is cause for optimism. In recent decades, gender equality has developed fairly rapidly, and if anything the pace of development only seems to be accelerating. One of the most visible indicators is the degree to which women have been entering the political arena, all the way to the top. With Golda Meir (Israel) and Indira Gandhi (India) as early forerunners, the 1980s and 1990s witnessed a small wave of female presidents and prime ministers, leading governments and countries all over the world. Among the names that come to mind are Margaret Thatcher (UK), Gro Harlem Brundtland (Norway), Cory Aquino (the Philippines), Benazir Bhutto (Pakistan), Tansu Çiller (Turkey) and Mary Robinson (Ireland).[12]

Megawathi Sukarnoputri (Indonesia) was one of the first female leaders of the twenty-first century, to be followed in recent years by Gloria Arroyo (Philippines), Julia Tymoshenko (Ukraine), Angela Merkel (Germany), Ellen Johnson-Sirleaf (Liberia), Michelle Bachelet (Chile), Portia Simpson Miller (Jamaica) and Han Myeong-Sook (South Korea).[13]

It is important to realize that the arrival of women in the political arena is no coincidence: a worldwide process of modernization is driving cultural change that encourages the rise of women in public life as well as the development of democratic institutions.[14] Of course, this is also a reminder that modernization is still spread unevenly, and that the position of women in many societies has not stopped being a weak and subordinate one.

One last note: while reaching top positions, not only in politics but also in business organizations, women might very well be able to display a new sort of leadership – feminine leadership – that is both 'tough and tender'.[15] Of course, the kind of difference this makes cannot be taken for granted and may need decades rather than years to really take effect. In the same vein, whether and to what degree female leadership would change the armed forces is not self-evident. Apart from that, however, it seems very unlikely that the military will not be influenced more and more by the general developments we have just depicted. In fact, as we will now argue, these are some of the very reasons why the armed forces should go on to invest in 'diversity'.

Managing diversity

Diversity management in work organizations has been on the (reseach) agenda for about half a century.[16] More recently, these insights have also been applied to armed forces, setting off specialized research and literature in addition to the more politically oriented works we discussed above.[17] Generally speaking, the importance of addressing diversity seems rooted firmly in most armed forces, and the lessons from literature, as well as from experience, are learned more or less on a day-to-day basis. Still, it makes sense to formulate somewhat more precisely where the relevance of cultural diversity lies. Why, again, is it so important for armed forces to incorporate this issue in their human resources management? In our view, there are at least six major reasons, some of them echoing observations we have already made in this introductory chapter.

The first reason relates to what has famously been called by Castells 'the power of identity'.[18] Although it has often been said that 'the end of ideology' has arrived, civil movements are on the rise, including freshly formed or already established minority groups stressing and cultivating their collective identity. Increasingly they have begun to demand their rights of citizenship, for example with regard to education, the labour market and political organizations.[19] This demand does not exclude the armed forces. Women and ethnic minorities, whether immigrants or indigenous groups, as well as gays, lesbians and other identity groups, want to exert their right to enter any position in the military. They want to have a fair chance to reach the higher ranks as well as positions in technologically sophisticated branches and special operations units. In sum, empowered by identity politics, these groups are demanding basic civil rights, demands that the military cannot afford to resist.

Second, seen in a broader context, these rights and claims remind the armed forces of the necessity to create and preserve sufficient legitimacy among their stakeholders in society: politicians, the general public, churches, unions and employers, to list a number of the most important ones. Traditionally, the general belief has been that the military workforce should mirror the workforce in society, or – as in the case of conscription – the (young) male population at large.

By ensuring a more or less equal representation with respect to socio-economic class, political conviction, religion and region, the armed forces in many countries around the world were able to become national institutions *par excellence*. On the one hand, the representation of all parts of the population has been looked upon as a mechanism for controlling the armed forces. On the other hand, in this way the military have been able to fulfil a nation-building role by educating and 'nationalizing' the poorer, less developed people, including those who happened to live on the peripheries of their society. Thus, since the time of Napoleon the armed forces have functioned as 'the school of the nation'. Through their army experience, 'peasants became Frenchmen', as the famous saying goes.[20]

This development has been observed not only in Europe and North America, but also in countries as different as Turkey, India and Brazil. Nowadays, the armed forces, whether following the conscription or the professional format, can contribute to their legitimacy if they once again fulfil this role of national institution. Since minority groups in society at large are often relatively deprived in terms of education and job opportunities, the armed forces can help by positioning themselves as a vehicle for social mobility. By doing so, they strengthen their own legitimacy.

The third reason for paying serious attention to diversity is a consequence of the professionalization of armed forces. In more and more countries, all-volunteer forces have replaced conscript armies, a process that brings with it a number of serious challenges. The automatic influx of personnel, taken for granted under the conscript system, has been superseded by a laborious process of recruiting and contracting soldiers and officers. In many countries, more often than not it proves difficult to fill all the vacancies in the armed forces. In particular, the technical and front positions in the army, air force and navy are often hard to fill. Here lies what may be called the business case for managing diversity.

But the recruiting problem is even more challenging. The armed forces have to compete with other employers in society, which can often offer higher salaries, better career opportunities and, on top of that, a safer working environment. Especially if a country's military are frequently involved in dangerous, life-threatening operations, recruitment may become truly difficult.[21] Moreover, competition in the labour market is likely to become even fiercer at a time when the population in general and the workforce in particular are decreasing in absolute numbers. This is a development that is bound to happen in many Western societies.

Against this background, it is in the military's own interest to expand the pool of potential recruits as much as possible with new categories of personnel such as women, ethnic minority groups and even foreigners. As of 2003 there were approximately 30,000 foreign-born, non-US citizens in the American armed forces, including many Jamaicans, Mexicans and other Hispanics. For them, serving in the military is a route to becoming a US citizen. The latter is a bonus that contrasts with the way in which the British Army still relies on its tradition of recruiting significant numbers of non-UK citizens, especially Gurkhas and Fijians.[22]

The fourth reason for emphasizing the importance of recruiting and managing diverse armed forces has to do with improving the effectiveness of military units, especially when they are engaged in operations other than war. In the context of humanitarian relief missions and civil–military cooperation, intercultural awareness and likewise communicative skills are usually more important than competency as warriors.

On the basis of a study of US units in Somalia, Miller and Moskos[23] have suggested that female, as well as African-American, soldiers generally are better equipped than white males to perform humanitarian missions conducted in non-Western countries. The reason is that in these circumstances women and black men seem to show relatively more empathy with local people.

In a somewhat similar vein, Soeters and others have compared the Turkish and Dutch approach during military operations in Afghanistan.[24] It appeared that the Turkish soldiers – owing to their more refined cultural sensitivity in that particular part of the world – were better able to cope with problems occurring in the interaction with the local population. As a result, the Turks were seldom attacked. One could argue that this kind of intercultural competence, which strengthens perceived mission legitimacy, also belongs to the baggage of, for instance, Turkish-Dutch, Turkish-German and Turkish-Belgian soldiers. For the armed forces of these countries, this provides an extra incentive for recruiting ethnic minority soldiers.

To give yet another example, on the basis of a thorough evaluation of civil military projects in Kosovo and Afghanistan, Rietjens has argued that female military officers may contribute to the success of such projects, especially when interaction with female beneficiaries in the host nations is involved.[25]

Clearly, research findings like these have to be amplified with the results of more case studies in order to refine and solidify our insights in a scientific way. However, the general point that armed forces stand a better chance of success when to some degree their own diversity matches the culture and diversity of host nations is very plausible indeed.

This latter point may apply also in the context of home countries – and this is the fifth reason for the necessity of having diverse armed forces. When particular minority groups are hardly represented in the military workforce, soldiers may be prejudiced vis-à-vis those groups. This may lead to serious malfeasance when the military are called upon to take action in case of civil disturbances. For example, when, in 2003, the Bolivian armed forces, with virtually only officers of Spanish descent in their ranks, were ordered to control protest demonstrations by indigenous people in the streets of La Paz and Cochabamba, more than 100 protesters were killed. It is highly unlikely that the military would have acted in the same manner had the protesters been predominantly of European origin (see Chapter 4).

Similar incidents have been observed with respect to Indian police forces containing a vast majority of non-Muslims, actively participating in indiscriminate shooting, the majority of their victims being Muslims.[26] Interestingly, as is argued in Chapter 8, the Indian military may be better equipped to deal with civil unrest in an impartial way than the police.

The sixth reason underscoring the importance of managing diversity in the armed forces, applies when, after a political turnaround, previously opposed groups have to be integrated in order to build a new army. A recent example of such a process in Europe can be found in the integration of the Bundeswehr and the Nationale Volksarmee after the unification of West and East Germany in 1990.[27] In other parts of the world this situation occurs more frequently, with South Africa as a prime example. Chapter 6 testifies to the urgency and the complexity of merging different armed forces in this particular circumstances. Of course, in a situation like that, managing diversity is no longer a matter of choice or of enlightened policies. It is, rather, a necessity in a completely new situation, unavoidably creating frictions and tensions among the people and the groups involved.[28] In those circumstances a number of insights from the literature on diversity management have a special relevance. In the next section we will briefly draw attention to those particular insights.

Uncomfortable issues

Shortly after the beginning of the operations in Afghanistan, James Lee, a Chinese-American West Point graduate, became a Muslim chaplain in the US Army, working for detainees at Guantánamo. Because Lee occasionally interceded on behalf of prisoners who he considered were being treated in an unnecessarily provocative way by their guards, he began to attract suspicion. The fact that he regularly met with the forty or so Muslim servicemen on the base only enhanced suspicions, particularly in the eyes of his commanding general. On 10 September 2003, one day before the second anniversary of the terrorist attacks on the twin towers, Lee was taken into custody, charged with espionage, association with known terrorists, and later – when none of these charges turned out to be true – with adultery and downloading pornographic material on to his laptop. It took eight months before the US Army dropped all charges, but clearly the chaplain had no other choice than to leave the military.[29] While not unique, Lee's story is a dramatic incident reflecting the military's anxiety that is brought about in a time of increasing terrorism.

Since the beginning of the twenty-first century, the position of minority groups in Western societies – especially minority groups with a Muslim background – has worsened. The terrorist attacks on the twin towers, the bombings in Bali, Madrid and London, the assassination of Dutch filmmaker Theo van Gogh have all contributed to a climate of fear and antagonism. The war in Iraq has done little to ameliorate this climate. The upheaval about cartoons in a Danish newspaper, considered to be offensive by many Muslims, illustrated how local incidents can stir up emotions and create tensions within and between societies all over the world. Somewhat paradoxically, in such a global climate of fear, people tend to become less cosmopolitan and more inward-looking.[30]

It is only to be expected that developments like these make themselves felt throughout the key institutions of society, such as the judiciary, the school system and public housing. Of course, the labour market and workforces too are affected by cultural tensions. At the same time, all these institutions are part of the solution, by providing the fundamental mechanisms through which social and economic integration can occur. As we argued earlier, the military might very well be able to play a useful role by taking up part of their tradition as school of the nation.

However, like society in general, the armed forces are bound to be affected by intercultural tensions and thus by uncomfortable isues. The story of Captain Lee offers a clear illustration – though we certainly do not want to suggest that it is in any way typical of the position of Muslims within the US military. The point is, even if standard policies and practices fit in with sophisticated ways of diversity management, the tensions and patterns of society cannot simply be excluded.

To give another example, a survey of the Dutch armed forces showed a deterioration in the diversity climate over the past five years. Between 2000 and 2005 the attitudes of the Dutch service(wo)men towards colleagues of Dutch minority groups in the military have become less favourable. On average, the respondents developed a slightly more negative view than before.[31] This is hardly a dramatic setback, but the general tendency is clear enough. Interestingly, the respondents indicated that new policy rules and regulations are necessary to turn the tide and improve the diversity climate in the Netherlands armed forces.

This brings us to the general question of how to organize a diverse workforce in such a way as to avoid worsening labour relations in the units and instead to enhance unit performance and effectiveness. Of course, in the past the general idea was that 'diverse' personnel in military units should not be mixed. When in the Ottoman Empire a law on the conscription of non-Muslims was passed, the non-Muslim population demanded that their men should serve in separate units from Muslim troops.[32] In World War I, Canada deployed all-black units,[33] just as the United States did up to the end of World War II (see Chapter 2).

In the British Indian Army, many battalions and regiments consisted of companies that were ethnically homogeneous. In this way, a battalion might have one company of Sikhs, one of Punjabi Muslims, one of Pathans and one of Dogras.[34] Nowadays, in the highly diverse Indian armed forces this way of organizing manpower persists, because it has proved to be very useful (see Chapter 8).

However, in most modern armed forces such policies of segregation are generally perceived to be unacceptable, not only because they are considered unnecessary and ineffective, or simply impossible to implement, but also, and especially, because they would contradict the ideals of multicultural integration, common citizenship and equal rights. Nevertheless, as will become clear in this book, in some other countries besides India, modicums of separation are still in vogue, either for political or for operational reasons.[35]

All in all, the historical mainstream of managing diversity has been away from separation. As we have already noted, this applies to gender as well. Even though in this case the trend is rather more recent, in most countries the phenomenon of the separate women's corps has disappeared.

So today mixed work units are the norm, even though reviews of research suggest they can be a mixed blessing. One way or another, in order to become a positive force for enhancing effectiveness, diversity needs sustained attention and careful study. So far it has become clear that, depending on a number of variables, group diversity can show either positive or negative effects.[36]

Generally speaking, three variables have emerged as crucial in the relationship between diversity and performance: the nature of the group's task, the numerical composition of the group, and the work climate, including the group's leadership. A few remarks on these three variables may be helpful. First, tasks that are computational, on the basis of technology-driven instructions and standard operating procedures (such as are formed in air forces), are less likely to produce diversity problems. The same holds true for the opposite kinds of tasks, those of a non-routine nature, as in humanitarian aid or R&D. However, other tasks – in particular those that aim at performing coordination and implementation activities – have a bigger chance of being jeopardized by unit diversity.

Second, teams or units that are either homogeneous with only a limited number of 'exceptions' or – on the other hand – units that are fully heterogeneous give rise to fewer problems than moderately heterogeneous groups. Within the latter, equally sized groups are more likely to start politicking and, hence, endangering the group's performance.[37] Finally, if group leadership is strong, motivating group members to strive for the achievement of the group's goals, stressing equal status, taking responsibility during incidents and displaying a certain 'diversity awareness', problems are less likely to develop.

Of course, these insights with respect to the relation between diversity and unit performance could be elaborated more fully and are in need of more rigorous scientific proof. However, the general patterns are clear. This also applies to what has become known as the 'token' position of minorities, which now and then creates highly uncomfortable issues in the military. It can also be a general manifestation of the difficult position members of minority groups find themselves in. If they are on their own or form a relatively small group in an organization's unit, minority members are 'tokens', as sociologist Rosabeth Moss Kanter has phrased it in her work on gender relations.[38] Tokens attract attention because of their high visibility. They capture a larger share of awareness, and they experience more performance pressure because they are always in the limelight. In this situation they need to toe the fine line between doing just well enough and doing too well. Furthermore, tokens are subject to the tendency to perceive or exaggerate differences, which can easily lead to their isolation because the dominant group boundaries are emphasized. Tokens can only break their isolation if they regularly pass informal loyalty tests. Finally, tokens experience more stereotyping and generalizing about the (cultural) group they belong to. In such a situation it is difficult to do things right.

An example of this process is offered by the story of a young female Turkish-Dutch corporal. While she was on a mission in Bosnia, her commander had asked her to liaise between the Dutch and a neighbouring Turkish battalion in order for them to develop good working relations. Because of her bilingual skills she was highly competent to perform this liaison role. After she had enthusiastically began to do what she had been asked to, she gradually came under criticism from both sides: the Dutch thought she was meeting the Turks too often, and the Turks found her behaviour and attitude too Dutch.

The ensuing tension prevented her from pursuing her military career in the Dutch armed forces, upon which she decided to continue her life as a civilian student. Clearly, she is not the only example of members of minority groups in militaries across the world who have experienced, or are experiencing, serious troubles while – or immediately after – performing more than adequately in their soldier's job. These stories not only demonstrate that members of minority groups often find themselves in truly difficult positions but also underscore the fact that diversity is often a mixed blessing and requires careful and sustained attention on the part of all the people involved.

Fourteen countries

Following this introductory chapter this book contains thirteen chapters on cultural diversity in the armed forces of fourteen different countries. We will venture on a global *tour d'horizon*, starting in the Americas, North and South, going from the United States and Canada towards Bolivia, Ecuador and Brazil. Thereafter we cross the Atlantic, landing in South Africa and hopping towards Eritrea. Then we go east to deal with India and then Israel. Our tour is rounded off by visiting a handful of European countries: the United Kingdom, France, Germany, Belgium and the Netherlands. There are no doubt many other countries in which the issue of cultural diversity is a relevant or even a dramatic one, but even as it stands, the countries presented in this book display a fascinating spectrum. All of them have their own stories and histories with respect to diversity, often problematic, sometimes painful, sometimes successful. To these stories and histories this book is devoted.

In all fourteen country cases, ethnic diversity will be addressed. With regard to three countries (Bolivia, Ecuador and the United Kingdom) the focus is exclusively on ethnic diversity. For the other eleven countries, gender is the second leading example of diversity. In the majority of the chapters diversity is also analysed in terms of sexual orientation. From the 1980s onwards the position of homosexual men and women in the military has been the subject of debate and research.[39] In terms of civil rights and identity politics, there have been and still are clear parallels with the rise of women in the military. On the other hand, though, gays and lesbians very much had and have to fight their own struggle in order to officially gain access to the armed forces. While they have been generally successful in doing so, often their day-to-day life as soldiers and officers is far from easy and not free from prejudice and discrimination. Even in countries where 'don't ask, don't tell' is not the official policy, it can be looked upon by the parties involved as the preferred strategy for coping.

Depending on national circumstances, other kinds of diversity will be on the table as well: language, religion, class, caste and region. Of course, sometimes these variables tend to cluster, ethnicity included. The ways in which all these kinds of diversity are defined and dealt with differ from country to country, both because their salience, in terms of prominence and debate, varies, and because diversity as a social category (or social construction) can have different connotations and implications in different places. Moreover, diverse groups are not homogeneous themselves, and to recognize this fits in with the thrust of diversity management as it has been defined in the leading literature: it is not only about 'recruiting and retaining employees from different demographic backgrounds', but especially about 'using their qualities in an optimal way'.[40] This presupposes an awareness of subtle demographic and cultural distinctions, an awareness that goes well beyond the gross statistics of representativeness.

Evidently, all fourteen countries and their armed forces vary with respect to notable features such as geopolitical position, operational experience and the character of recruitment (conscription versus all-volunteer force). As will become clear also, often in the armed forces themselves the different services and branches display different records vis-à-vis diversity. Within the organizational archipelago of the armed forces, the representation of women as well as ethnic minorities, can be very uneven indeed.

All this contributes to the diversity of diversity. On top of that, the authors of the chapters add their own bit of diversity, as a result of the use and availability of data, of different academic backgrounds, but also because of differences in viewpoint and appreciation. Surely, academia does not function in a cultural vacuum, and making solid scientific assessments does not preclude the drawing of polemical conclusions. The first country chapter, that on the United States, immediately offers a provocative case in point. In that chapter, Charles Moskos concludes that 'we cannot expect to have truly succesful race relations and gender relations in the same organization'. As Moskos himself is no doubt aware, not everybody will agree with this statement. One way or another, however, it offers a challenging hypothesis about the interrelationships of different kinds of diversity within the military.

This book offers a wealth of data and insights for testing and refining the plausibility of Moskos's hypothesis. But, of course, this is only one example of how comparing different chapters and cases can be fruitful and stimulating. We would hesitate to argue that there is a single pattern underlying the development of diversity in the armed forces. But we are strongly convinced that for military establishments, political leaders and civil stakeholders in all countries involved, there is much to be learned from each other's best practices. We hope and expect that this also holds true for the many countries not represented in this book, because, deep down, cultural diversity poses the same challenges to societies and armed forces all over the world.

Notes

1 *De Volkskrant*, 16 January 2006.
2 P.L. Berger and S.P. Huntington (eds), *Many Globalizations: Cultural Diversity in the Contemporary World*, Oxford: Oxford University Press, 2002.
3 See, for instance, for a strongly argued case against the dangers of diversity: 'Culture wars within the military', *Orbis*, 43, 1999, 9–57.
4 J. Jimmerson, 'Photo Book Celebrates Tuskegee Airmen', *Post-Herald-Albama*, 15 February 1999; The Eleanor Roosevelt Papers, 'The Tuskegee Airmen', in A. Black, J. Hopkins *et al.*, *Teaching Eleanor Roosevelt*, Hyde Park, NY: Eleanor Roosevelt National Historic Site, http://www.nps.gov/elro/glossary/tuskegee-airmen.htm (accessed 14 April 2006).
5 C. Finkel, *Osman's Dream: The History of the Ottoman Empire*, New York: Basic Books, 2005, pp. 470–471.
6 C.H. Enloe, *Ethnic Soldiers: State Security in Divided Societies*, Athens: University of Georgia Press, 1980.
7 N.F. Dreisziger (ed.), *Ethnic Armies: Polyethnic Armed Forces. From the Time of the Habsburgs to the Age of the Superpowers*, Waterloo, Ontario: Wilfrid Laurier University Press, 1990.
8 A. Peled, *A Question of Loyalty: Military Manpower Policy in Multiethnic States*, Ithaca, NY: Cornell Unversity Press, 1998.
9 Ibid., pp. 1–26.
10 For a concise overview of debates and data, including a bibliography, see M.L. Kendrigan, 'Gender and the warrior ethic', paper given at the bi-annual conference of the Inter University Seminar on Armed Forces and Sociey, Chicago, October 2005.
11 For a comparative study on Australia, Mexico and Zimbabwe, see D. Iskra, S. Trainor, M. Lihauser and M. Wechsler Segal, 'Women's participation in armed forces cross-nationally: expanding Segal's model', *Current Sociology*, September 2002, 771–797.
12 Golda Meir was prime minister from 1969 to 1974. Indira Gandhi from 1966 to 1977 and from 1980 till her violent death in 1984.
13 See Trudy Rubin, 'Will Women Show a New Way?', *Current History*, 105, 689, March 2005, 100–105.
14 R. Inglehart, P. Norris and C. Welzel, 'Gender equality and democracy' (mimeo), Ann Arbor: University of Michigan, 2006.
15 R. Buddan, 'Defining Portia: Masculine and Feminine Leadership', *Jamaica Gleaner*, 19 March 2006; see also S. Helgesen, *The Female Advantage: Women's Ways of Leadership*, New York: Doubleday, 1990.
16 K.Y. Williams and C. O'Reilly III, 'Demography and Diversity in Organizations: A Review of 40 Years of Research', *Research in Organizational Behavior*, 20, 1998, 77–140.
17 For a good example, see M.R. Dansby, J.B. Stewart and S.C. Webb (eds), *Managing Diversity in the Military: Research Perspectives from the Defense Equal Opportunity Management Institute*, New Brunswick, NJ: Transaction Publishers, 2001.

18 M. Castells, *The Power of Identity: The Information Age: Economy, Society and Culture*, vol. 2, Oxford: Blackwell, 1997.
19 W. Kymlicka, *Politics in the Vernacular: Nationalism, Multiculturalism and Citizenship*, Oxford: Oxford University Press, 2001.
20 E. Weber, *Peasants into Frenchmen: The Modernization of Rural France, 1870–1914*, Stanford, CA: Stanford University Press, 1976. For a critical reappraisal of the military as school of the nation, see R.R. Krebs, 'A School for the Nation? How Military Service Does Not Build Nations, and How It Might', *International Security*, 28, 4, Spring 2004, 85–124.
21 J. van der Meulen and J. Soeters (eds), 'Considering Casualties', special issue of *Armed Forces and Society*, 31, 4, Summer 2005.
22 See T. Barkawi, *Globalization and War*, Lanham, MD: Rowman & Littlefield, 2006, pp. 46–47.
23 L.L. Miller and C. Moskos, 'Humanitarians or Warriors? Race, Gender and Combat Status in Operation Restore Hope', *Armed Forces and Society*, 21, 4, Summer 1995, 615–637.
24 J. Soeters, E. Tanercan, A. Varoglu and U. Sigri, 'Turkish–Dutch Encounters in Peace Operations', *International Peacekeeping*, 11, 2, 2004, 354–368.
25 S.J.H. Rietjens, 'Civil–Military Cooperation in Response to a Complex Emergency: Just Another Drill?' Ph.D. thesis, University of Twente, the Netherlands, 2006, p. 199.
26 J. Soeters, *Ethnic Conflict and Terrorism: The Origins and Dynamics of Civil Wars*, London: Routledge, 2005.
27 See N. Leonhard, 'Armee der Einheit: Zur Integration von NVA-Soldaten in die Bundeswehr' (Army of unity: about the integration of NVA soldiers in the Bundeswehr), in S. Bernhard Gareis and P. Klein (eds), *Handbuch Militär- und Sozialwissenschaft* (Handbook of Military Social Science), Wiesbaden: VS Verlag, 2004, pp. 70–80. For the integration of the Berlin police, see A. Glaeser, *Divided in Unity: Identity, Germany, and the Berlin Police*, Chicago: University of Chicago Press, 1999.
28 See I. Liebenberg, 'The Integration of the Military in Post-Liberation South Africa: The Contribution of Revolutionary Armies', *Armed Forces and Society*, 24, 1, 1997, 105–132.
29 J. Lee and A. Molloy, *For God and Country: Faith and Patriotism under Fire*, New York: Public Affairs, 2005.
30 M.R. Olivas-Lujan, A.-W. Harzing and S. McCoy, 'September 11, 2001: Two Quasi-experiments on the Influence of Threats on Cultural Values and Cosmopolitanism', *International Journal of Cross-Cultural Management*, 4, 2, 2004, 211–228.
31 F. Bosman, R. Richardson and J. Soeters, 'Intercultural Tensions in the Military: Evidence from the Netherlands Armed Forces', 2006 (manuscript under review).
32 C. Moskos and J. Sibly Butler, *All That We Can Be: Black Leadership and Racial Integration*, New York: Basic Books, 1996; Finkel, *Osman's Dream*, p. 159.
33 John G. Armstrong, 'The Unwelcome Sacrifice: A Black Unit in the Canadian Expeditionary Force, 1917–19', in Dreisziger (ed.), *Ethnic Armies*, pp. 178–197.
34 Barkawi, *Globalization and War*, p. 72.
35 Separation by way of informal self-selection is yet another phenomenon. Research in the US Army has demonstrated that – and why – minorities are poorly represented in special operations forces. One of the factors was the perception that army special forces are white organizations with racist attitudes. This perception is shared by both military and civilian communities. This perception deters minorities – especially African-Americans – from considering careers in such forces. As the researchers indicate, the US Army should determine whether a basis for perceptions of racism exists. Obviously, this is a rather uncomfortable issue, and one that is not likely to be limited to the US forces only. See S.N. Kirby, M.C. Harrell and J. Sloan, 'Why Don't Minorities Join Special Operations Forces?', *Armed Forces and Society*, 26, 4, 2000, 523–545.

36 Williams and O'Reilly, 'Demography and Diversity', p. 120; D. Van Knippenberg, C.K.W. de Dreu and A.C. Homan, 'Work Group Diversity and Group Performance: An Integrative Model and Research Agenda', *Journal of Applied Psychology*, 89, 6, 2004, 1008–1022.
37 For a review of the impact of these factors with respect to international military cooperation (which of course is also a form of diversity management), see J. Soeters, D. Resteigne, R. Moelker and P. Manigart, 'Smooth and Strained International Military Cooperation', 2006 (manuscript under review).
38 R.M. Kanter, 'Some effects of group proportions in group life: skewed sex ratios and responses to token women', *American Journal of Sociology*, 82, 1977, 965–990. See also R.M. Kanter, *Men and Women of the Corporation*, New York: Basic Books, 1977.
39 W.J. Scott and S. Carson Stanley (eds), *Gays and Lesbians in the Military*, New York: Aldine de Gruyter, 1994; G.M. Herek, J.B. Jobe and R.M. Carney (eds), *Out in Force: Sexual Orientation and the Military*, Chicago: University of Chicago Press, 1996.
40 This is a paraphrase of a definition of diversity management by P. Prasad, A.J. Mills, M. Elmes and A. Prasad, *Managing the Organizational Melting Pot: Dilemmas of Workplace Diversity*, Thousand Oaks, CA: Sage, 1997.

2 Diversity in the armed forces of the United States

Charles Moskos

To describe the United States as a country of diverse ethnic groups is to state the obvious. That the US military has been one of the most successful institutions for black achievement and race relations is also generally acknowledged. More recently, the incorporation of women into non-traditional military roles has also proceeded apace. Gender integration, however, has not been as complete or as smooth as that of race. The differences and similarities between race and gender relations in the US armed forces are our focus here. The plan of the chapter is straightforward. First, we look at race. Second, we examine gender. Third, and finally, we look at race and gender together.

Race

This is not the place to give a full history of blacks in the military, but some background may be useful. African-Americans have served in America's armed forces from colonial times.[1] With only a few exceptions, such service was characterized by racial segregation (though with white officers) through World War II.

A major threshold was crossed in the winter of 1944-45. The Battle of the Bulge had severely depleted infantry reserves. Searching desperately for new sources of replacements, the Army asked for volunteers from black support units in the theater. Over 4,500 black soldiers answered the call, 2,500 of whom were accepted. The black soldiers' excellent combat performance, coupled with a lack of serious friction from white troops, made the experiment an unqualified success. Significantly, survey data showed that the more contact white soldiers had with black troops, the more favorable was their reaction toward racial integration. This became one of the landmark findings of *The American Soldier* and was used to buttress later desegregation decisions.[2]

In another precedent-breaking experiment on the race relations front, the all-black 332nd Fighter Group, trained at Tuskegee, Alabama, engaged in aerial combat over Italy and escorted bombers deep into Germany. Not one bomber entrusted to the 332nd was lost to enemy fighters – a claim no other unit could make. The 'Tuskegee Airmen', whose flying crews were black commissioned officers, proved that black aircrews were at least as capable as white ones.

The 332nd was commanded by Benjamin O. Davis, Jr., who would later become the first black to hold three-star rank. Nearly three decades after retiring from a military career spent fighting enemies abroad and racial barriers at home, General Davis was awarded a fourth star. In a public ceremony in December 1998, President Clinton pinned the military's highest peacetime rank on the 88-year-old former general.

The first step toward desegregation of the armed forces was an executive order issued by President Truman in 1948. Racial integration really began with the Korean War and was accomplished *de facto* by the early 1950s. The period between the wars in Korea and Vietnam was one of relative racial calm in the armed forces. Over the course of the Vietnam War, however, white–black polarization heightened, with racial clashes occurring worldwide throughout the services. Racial conflict did not disappear in 1973 with the end of the draft and the Vietnam War. In many ways it grew worse. Fights between black and white soldiers were endemic in the 1970s, an era now remembered as the 'time of troubles'.

Since the early 1980s, race relations have generally been positive in the armed forces. Much of this is due to equal opportunity advisers (EOAs), who monitor racial incidents and look at patterns of race in assignments and promotions. The EOAs are trained at the Defense Equal Opportunity Management Institute (DEOMI). When founded in 1971, the organization was called the Defense Race Relations Institute (DRRI). That name reflected its original purpose: to cope with the racial turbulence then afflicting the military. With the increasing number of women in the armed forces, the mission for equal opportunity broadened and led to the DEOMI designation in 1979. Indeed, the most significant difference between DRRI and DEOMI is its shift from issues of racial discrimination to gender issues and sexual harassment.

Today, give or take a surly remark here, a bruised sensibility there, the races get on well. A rule of the thumb is that the more military the environment, the more complete the integration. Interracial comity is stronger in the field than in garrison, stronger on duty than off, stronger on post than in the world beyond the base. In surveys conducted in Somalia, Bosnia and Iraq, over three-quarters of both white and black soldiers say that race relations are better in the Army than in civilian life.

As shown in Table 2.1, the proportion of blacks in the US armed forces has risen substantially since the end of World War II. Blacks in 2004 comprised 20 percent of the armed forces, compared to about 13 percent of the American population. With the end of conscription in 1973, the proportion of blacks in the military increased significantly. Since 2000, and especially since the start of the Iraq War in 2003, however, the black percentage has dropped somewhat, owing to a decline in black enlistments (though not re-enlistments). The cause for this drop has been variously attributed to disenchantment with the George W. Bush administration and increasing black opportunities in the civilian sector as well as the war in Iraq.[3]

Contrary to conventional wisdom, blacks did not suffer disproportionate

Table 2.1 Blacks as a percentage of US armed forces, selected years 1949–2004

Year	Enlisted	Officers	Total
1949	8	1	7
1964	10	2	9
1970	11	2	10
1980	22	5	19
1990	23	7	20
1998	22	8	20
2004	20	9	18

Source: Department of Defense.

casualties in the wars in Vietnam, Afghanistan or Iraq. As of September 2005, blacks, who comprise 18 percent of all military members, accounted for 9 percent of those killed in Iraq. Hispanics, who comprise 9 percent of the force, accounted for 9 percent of those killed. In reality, the number of casualties suffered in the wars in both Iraq and Afghanistan have been disproportionately high for working-class whites coming largely from rural areas[4] – a fact that has not caused concern in the American media.

The black composition varies significantly by service and rank, as shown in Table 2.2. We focus on the Army because it is the largest of the services and the one with the highest proportion of blacks. In 2005 the 115,000 blacks in the Army constituted half of all blacks in military uniform. By rank, the number of blacks in the Army breaks down as 16 percent of the lower enlisted levels, 36 percent of non-commissioned officers (NCOS) and 12 percent of commissioned officers. By 2005 there were proportionately twice as many black senior NCOs as lower-ranking enlisted members. Also noteworthy is that 8 percent of all Army generals are black, a figure double the proportion of flag rank officers in the other services. A most significant event was the 1989 appointment of Colin L. Powell, the son of Jamaican immigrants, to Chairman of the Joint Chiefs of Staff, the most powerful military position in the world.

In noting the success of race relations in the Army, we do not intend to turn a blind eye to real and serious problems that continue to exist. Certainly the Army is not a racial utopia. Black and white soldiers are susceptible to the same kinds of interracial suspicion and resentment that exist outside the Army. Though it stands in favorable contrast to non-military institutions, the Army is not immune to the demons that haunt race relations in the United States.

Whatever its racial tensions, however, the Army stands out as an organization in which blacks succeed, and often surpass their white counterparts. The armed forces are the only place in American society where whites are routinely bossed around by blacks.

Table 2.2 Blacks as a percentage of total personnel by grade and service, 2004

Grade[a]	Army	Navy	Air Force	Marine Corps
Commissioned officers				
O-7+ (generals)	8	3	4	5
O-6 (colonel)	9	4	6	4
O-5 (lt. Colonel)	12	4	5	4
O-4 (major)	12	8	6	6
O-3 (captain)	12	8	7	8
O-2 (1st lieutenant)	14	9	8	5
O-1 (2nd lieutenant)	12	8	6	3
Total officers	12	7	7	6
Enlisted personnel				
E-9 (sgt. major)	41	10	17	32
E-8 (master sgt.)	36	13	20	28
E-7 (sgt. 1st class)	37	17	21	23
E-6 (staff sgt.)	33	20	18	19
E-5 (sergeant)	28	22	17	16
E-4 (corporal/specialist)	21	20	18	11
E-3 (pvt. 1st class)	16	23	15	9
E-2 (private)	16	21	16	8
E-1 (recruit)	15	24	13	8
Total enlisted	25	21	17	13
Total personnel	23	19	15	12

Source: Department of Defense.

Note
a Army titles in parentheses have equivalent ranks in the other services.

Lessons

Can any lessons be drawn for civilian life or other militaries in advanced Western democracies? I suggest a broad principle: race relations can best be transformed by an absolute commitment to non-discrimination coupled with uncompromising standards of performance. To maintain standards, however, paths of opportunity must be created – through education, training and mentoring – for those who otherwise would be at a disadvantage. Comprehension of how race relations work in the United States suggests five concrete lessons.[5]

Lesson 1: blacks and whites will not view opportunities and race relations in the same way

Even in the Army, the most successfully racially integrated institution in American society, blacks and whites still have disparate views of equal opportunity. Blacks consistently view racial matters in a less favorable light than do whites. This cuts across gender and rank. There is no foreseeable situation in any American institution, much less society as a whole, where this is likely to change. The aftermath of the 2005 disaster caused by Hurricane Katrina in New Orleans is a recent example of this conflicting view of current events. Black Americans were four times more likely (82 percent) than whites (19 percent) to attribute the mishandling of the relief operation to racism.[6]

What the Army does show, however, is that black and white social attitudes can become significantly closer in egalitarian settings with shared experiences. It also shows that blacks and whites do not have to hold identical views on the racial situation in order to succeed together.

Lesson 2: focus on black opportunity channels rather than eradicating racism

Better to have blacks in leadership positions in an organization with some white racists than to have an organization with few black leaders where openly racist bigots are absent. The proclivity in civilian organizations, notably the university world, is to foster a better racial climate though eradication of racist statements and symbols. Such efforts are meaningful only when accompanied by concrete steps to expand the pool of qualified students and faculty.

The presence, even if not in any great number, of white racist 'skinheads' in the Army points to a profound and counter-intuitive lesson. Lamentable as the presence of white racists may be, it is not the core issue. African-American history is eloquent testimony to how black accomplishment can occur despite pervasive white racism. In no way should the absence of white racists be considered a precondition for African-American achievement. This is one of the most significant morals of the Army experience.

In practical terms the Army has developed an affirmative action program based on 'supply'. This contrasts with the 'demand' – and more typical – version of affirmative action in which goals and quotas are established without prior efforts to enlarge the pool of qualified people. The goals in the Army promotion process are based not on the number of minority members in the Army, but on the number of minority members in the pool of potential promotees to the next higher rank. In simple terms, enough qualified minorities must be present in the promotion pool to make affirmative action work well.

Lesson 3: be ruthless against discrimination

Although formal efforts to prohibit racist expressions can be a way to avoid a genuine opening up of channels for black advancement, this does not imply a retreat from anti-discrimination. Racist behavior cannot be tolerated within the leadership of an organization. Individuals who display such tendencies must not be promoted to positions of responsibility. Racist behavior in the Army effectively terminates one's career. That one rarely hears racial remarks among Army NCOs and officers, even in all-white groups, reflects how much this norm is adhered to.

Lesson 4: a level playing field is not enough

The Army shows how youths with deficient backgrounds can meet demanding academic as well as physical standards. The Army has successfully introduced internal programs to bring young people up to enlistment standards, to raise enlisted soldiers to NCO standards, and to raise high school graduates to West Point admission standards. These programs are not targeted exclusively to minority soldiers, but the participants are disproportionately African Americans.

The Army's success in producing black leaders occurs because it recognized that a level playing field is not enough. And this points to another Army lesson for civilian society. Rather than compromised standards, soldiers are raised to meet competitive standards. Toward this end, the Army has established far-reaching educational programs that emphasize core academic skills.

At the enlisted level this involves one of the largest continuing education programs in the world. In 2004 some 40,000 soldiers enrolled in Functional Academic Skills Training (FAST), of whom 50 percent were black (and 20 percent Hispanic). The lessons focus solely on mathematics, reading and writing. Without FAST, the strong minority representation in the NCO corps would be impossible.

The US Military Academy at West Point remains the most prestigious source of commissions in the Army. Almost half of the black cadets who enter West Point are products of one of the most unusual secondary schools in America: the US Military Academy Preparatory School (USMAPS). The ten-month program, in effect a thirteenth year of high school, emphasizes academic competency in reading, writing and mathematics. Without USMAPS the number of black cadets would be perilously low. Remarkably, black 'prepsters' are just as likely (74 percent) to graduate from West Point as white direct admittees (75 percent). The Air Force and the Navy (which includes the Marine Corps) have USMAPS equivalents.

Evaluating programs designed to boost academic skills and test scores is fraught with difficulty. But there is sufficient evidence from the Army experience to show that 'intelligence' – as measured by achievement tests – can be raised significantly through programs that are well staffed, have motivated participants and adopt a military regimen.[7] Residential programs away from the participant's home area seem to be the most effective way to resocialize young people toward productive goals. As is done in the Army, such programs should emphasize mathematics, reading and writing.

If liberals must learn that white racism is not the central point, conservatives have something to learn too. The skill-boosting programs that produce so many black leaders in the Army are costly. Such programs require a big commitment of money and resources.

Lesson 5: enhancing black participation is good for organizational effectiveness

The blunt truth is that, the way most Americans see it, the greater the black proportion in an organization, the poorer its effectiveness. The armed forces are the welcome exception. That the disproportionately black Army stands out as one of the most respected organizations in American society has profound meaning. Not only have the military played a central role as an avenue of black achievement, but they have also shown that a large African-American presence has been conducive for the smooth operation of a major American institution. Ultimately, any program of engineered race relations must meet a single test. Does it improve the performance of the organization in which it is implemented?

Gender

As with race, space again prohibits an extended discussion of the history of women in the US armed forces.[8] Suffice it to say that prior to World War II, the role of women in the military was extremely limited. In World War II some 350,000 women served in the armed forces in separate corps. Most held jobs in 'traditional' women's work in healthcare and administration. The Women's Armed Service Act of 1948 gave permanent status to military women, but with the proviso that there would be a 2 percent ceiling on the proportion of women in the services (excluding nurses). No female generals or admirals were permitted. For the next two decades women averaged only a little over 1 percent of the armed forces. As shown in Table 2.3, the number of women has increased dramatically in recent decades. Whereas women (many officer nurses) in 1970 made up only 2 percent of the armed forces, by 2004 women comprised 15 percent of both the enlisted force and officer corps.

With the end of conscription in 1973, a series of barriers for women in the military fell in relatively rapid succession. Women entered officer training programs on college campuses in 1972 and were admitted to the military academies in 1976. Two years later the separate women's corps was abolished. In recent years, more and more positions have been opened to women. By 2000, women were serving aboard all warships (excluding submarines), in all flying roles (including combat aircraft, albeit in small numbers) and in all ground positions except those involving direct combat. In the contemporary period, basic military training is gender integrated in the Navy Air Force and in non-combat arms of the Army. Basic training remains gender separated in the Marine Corps and in the combat arms of the Army.

Table 2.3 Females as a percentage of US armed forces, selected years 1970–2004

Year	Enlisted	Officers	Total
1970	1	4	2
1980	8	7	8
1990	13	11	12
1998	16	14	16
2004	15	15	15

Source: Department of Defense.

Despite ongoing reports and commissions set up to look at gender relations, the status of women in the US armed forces in the early years of the twenty-first century had seemingly come to a kind of stabilization.[9] The probability of opening the ground combat arms to women is remote. It seems equally unlikely that women would be removed from serving aboard warships or as pilots in warplanes. No major changes in the training of men and women appear to be in the offing. However, a 2002 survey of university students revealed that mixed-gender basic training has a negative effect on enlistment propensity.[10]

One significant shift, however, dealt with 'fraternization', especially that between men and women. Up to 1999, dating was allowed between officers and enlisted persons in the Army if the parties were not in the same chain of command. Starting in 2000, the Department of Defense decreed that there would be an absolute ban on dating between officers and enlisted persons. In this regard, the Army would follow the fraternization rules already existing in the other services.

Table 2.4 gives the participation of women by service and rank in 2004. The number of women was highest in the Air Force (20 percent), followed by the Army (15 percent), and the Navy (14 percent), and was lowest in the Marine

Table 2.4 Females as a percentage of total personnel by grade and service, 2004

Grade[a]	Army	Navy	Air Force	Marine Corps
Commissioned officers				
O-7+ (generals)	4	5	5	4
O-6 (colonel)	12	11	12	3
O-5 (lt. colonel)	13	12	13	3
O-4 (major)	14	15	15	2
O-3 (captain)	18	16	21	6
O-2 (1st lieutenant)	21	17	22	9
O-1 (2nd lieutenant)	21	17	22	9
Total officers	17	15	18	6
Enlisted personnel				
E-9 (sgt. major)	9	5	5	3
E-8 (master sgt.)	10	6	11	5
E-7 (sgt. 1st class)	11	8	12	5
E-6 (staff sgt.)	13	10	15	6
E-5 (sergeant)	15	11	22	7
E-4 (corporal/specialist)	16	18	23	6
E-3 (pvt. 1st class)	16	18	22	5
E-2 (private)	15	15	28	7
E-1 (recruit)	14	15	20	5
Total enlisted	15	14	20	6
Total personnel	15	14	20	6

Source: Department of Defense.

Note
a Army titles in parentheses have equivalent ranks in the other services.

Corps (6 percent). Women service members were disproportionately represented in the lower enlisted and lower officer levels in all four services.

If race relations in the military are relatively good in comparison with civilian society, the ongoing difficulties of the services in gender relations have been the cause of ongoing embarrassment. In 1991 a major scandal occurred at the convention of Navy fighter pilots in Las Vegas, Nevada. Male pilots assaulted female Navy personnel and brought national shame to the Navy.[11] In 1996, incidents of male sergeants sexually harassing female recruits in training camps rocked the Army. Shortly thereafter, charges of sexual abuse were brought against an Army sergeant major (who was subsequently found not guilty). In 2005 a national commission was established to examine reports of widespread sexual harassment in the military academies.

Yet even these scandals could serve as an impetus for opening up the remaining proscribed roles for women. It had become an article of faith for feminist spokespersons to hold that sexual harassment would come under control only when women were no longer regarded as second-class members of the military – that is, no longer excluded from the combat arms. Certainly the 1991 scandal at the Navy pilots' convention facilitated the opening up of combat aircraft and warships to women.

Military sociologist Laura L. Miller has described the disjuncture between the feminist agenda and the concerns of the vast majority of military women. Rare is the enlisted woman who expresses a desire to enter the combat arms. But directly to the point, Miller's surveys show that only 2 percent of enlisted women believe sexual harassment would decrease if the combat arms were opened to women. In fact, 61 percent believe harassment would increase![12] (The rest thought it would not make much difference one way or the other.)

Gender relations in the military confront a fundamental dilemma. Do we want more female generals or less sexual harassment? Without women in combat arms, there will never be many female generals. But enlisted women state that harassment will increase if females are put into the combat arms. Further, surveys show that virtually all enlisted women say they do not want to be treated the same as men if that entails assignment to the combat arms. Acknowledging this dilemma should help clear the air.

As the war in Iraq proceeded, steps were taken by the Defense Department to assign women soldiers in support units to become integrated with combat units. This was countered by a US congressional bill in May 2005 specifically stating that the military must receive specific authorization from Congress before placing women closer to combat missions.[13] As of September 2005, forty-two women soldiers had died in Iraq (about 2 percent of all fatalities). The deaths of these women soldiers have not caused any major public concern.

There are also blind alleys in dealing with sexual harassment in the armed forces. One is to confound power relations with sex. If power relations are the key cause of sexual abuse in the military, then there are only two realistic options. Either reduce the authority of superiors, or separate men from women.

Also, what are we to make of survey data which show that enlisted women prefer male superiors?[14]

The other dead end is to blame the prevailing culture within the military. In a bizarre appointment the Secretary of the Army in 1997 hired as a temporary consultant an advocate of replacing the 'masculinist' culture of the military with an 'ungendered vision' of military culture.[15] For those who want to abolish the prevailing 'macho' culture, what is the preferred culture? Men don't tell dirty jokes in front of women? Men don't tell dirty jokes at all? Women laugh at dirty jokes when men tell them? Women tell dirty jokes to men?

Although issues of sexual harassment dominate much of the coverage of the media on gender relations in the military, these must be placed in context. A 2003 survey conducted by the author of soldiers serving in Iraq shows that a large majority of both men (74 percent) and women (85 percent) do not see gender-mixed units as causing major difficulties. The Iraq survey, as with earlier surveys conducted in Bosnia and Kosovo, also reveals that almost two out of three enlisted women and four out of five enlisted men have come to regard false sexual accusations as being at least as much of a problem as actual sexual harassment.[16] A 2004 survey conducted by the Defense Department Inspector General reported that 73 percent of women at the military academies perceived fraudulent complaints of harassment as a problem.[17] We are witnessing a movement from fear of reporting harassment to fear that men will be excessively punished. Paradoxically, female officers are much more likely to see sexual harassment as a problem than enlisted women.[18]

The issue of false accusations has become a new dynamic in gender relations. On sexual harassment, the overriding goal must be to get more reporting of genuine harassment cases and reduce the number of false accusations. False accusations undermine the credibility of true victims, who are now second-guessed. Male superiors are worried that their authority is being undermined because of potential sexual harassment charges. We must be wary of those who view the sexual harassment scandals simply being the result of a 'few bad apples'. The focus must be on organizational changes that directly address sexual harassment.

Recommendations

Some recommendations to insure more accurate reporting of genuine sexual harassment include the following.

Recommendation 1

A female in the Inspector General (IG) branch must be readily available so that an aggrieved woman can be assured of seeing her if she wants. Other options: (a) a female–male 'team' within the IG to screen all sex abuse cases; (b) or, more radically, an all-female chain of complaint within IG (as exists in Israel).[19]

Recommendation 2

Military law must reconsider the issue of 'constructive force' in which sexual activity is always rape in a superior–subordinate setting. The military's position that consensual sex between superiors and subordinates always involves an element of coercion flies in the face of common sense and real-world experience. In our rush not to blame the victim, we must be wary not to infantilize adult women.

Recommendation 3

Punishment of some sorts should be meted out to female subordinates who misbehave sexually. What are appropriate (even if differential) sanctions for both parties in such relations? Presently, the United States has a *de facto* gender-normed standard of morality which demeans women. Item: no female officer who misbehaved at the Tailhook convention was reprimanded. Item: regulations at an Army training base exempt female trainees from prosecution for soliciting sex with male drill sergeants.[20]

Recommendation 4

Addressing the ramifications of consensual sex among service members is a question that ought not to be avoided (as it presently is) any longer. How consensual sex within a unit affects cohesion needs honest investigation. Related to this is the impact of in-service adulterous affairs on the spouses of military members. Survey data indicate that female soldiers perceive wives of soldiers as being jealous of female soldiers.

One aspect of gender relations rarely discussed in media coverage of the military is the proportion of single females in the career military force. Table 2.5 presents the marital status by grade groupings in the US armed forces. In essence, the more the senior the female, the less likely she is to be married than her male counterpart. Reconciliation of a military career and family life impacts much more heavily on women than it does on men.

Table 2.5 Percent married, by rank and gender, 2004

Grade	Males	Females	M/F difference
O6 and above	94	66	+28
O4–O5	90	66	+24
O1–O3	59	45	+14
E7–E9	88	62	+26
E5–E6	71	55	+16
E4	40	43	−3
E1–E3	20	25	−5

Source: Department of Defense.

Race and gender

The intersect of race and gender requires special mention. As given in Table 2.6 the proportion of black females has increased markedly from 1980 to 2004. This is true for the armed forces overall, but especially so in the Army. African Americans account for 23 percent of all Army women officers. Most noteworthy, blacks constitute 42 percent of all Army enlisted women, which makes them the largest racial group among female soldiers.

One of the most sensitive issues in the US military is that in the most publicized cases of sexual harassment, a large majority of the accused were black male sergeants and a large majority of the accusers were white female recruits. Indeed, black civil rights organizations took note of this and raised questions as to the credibility of the accusers. At the least, anytime a cross-race sexual accusation is made, especially if it involves a black male and a white female, one must look at the evidence very carefully, more so than otherwise.

One way of looking at race and gender relations is to pose a hypothetical examination question. What best explains race/gender aspects of sexual harassment in the Army: (a) male black sergeants making sexual advances to female

Table 2.6 Percentage of female service members by race, 1980, 1998 and 2004, total military and army only

	1980	1998	2004
Officers			
Total military:			
White	85	76	68
Black	9	14	16
Other	6	10	16
Total	100	100	100
Army only:			
White	83	69	62
Black	11	20	23
Other	6	11	15
Total	100	100	100
Enlisted			
Total military:			
White	70	53	48
Black	25	35	32
Other	5	12	20
Total	100	100	100
Army only:			
White	57	40	38
Black	37	47	42
Other	6	13	20
Total	100	100	100

Source: Department of Defense.

white trainees; (b) white women flirting with black superiors; (c) black women knowing how to fend off harassment better than white women; (d) cross-race sexual harassment being more likely to be reported than same-race harassment? The correct answer is probably (e): all of the above.

Also noteworthy is that the race/gender effect on attrition (i.e. those who fail to complete their initial enlistment for one cause or another) in the contemporary military. For males of any race and for black females, about one in three do not complete their enlistment. (Note: the attrition rate for males in the all-volunteer force is three times greater than it was for draftees in the Cold War era.) For white women, however, the attrition rate is over one a half times higher than for men or black women. Indeed, in the Army, less than half of white women complete their initial enlistment!

Those seeking greater opportunity for women in the military typically draw analogies between women and African Americans. There are indeed similarities between race and gender equity in the Army. Both blacks and women are a minority of Army personnel (23 percent and 15 percent, respectively). Blacks served in segregated units, and did so until the early years of the Korean War. Women served in separate all-female units in World War II and continued to do so until the mid-1970s. And, to be sure, some of the current arguments that gender integration undermines unit cohesion are similar to those used by opponents of racial desegregation in the late 1940s.

But these apparent similarities must not obscure the fact that the situations of blacks and women in the military are not comparable. The contrasts between race and gender are summarized in Table 2.7.

Let us start with the most obvious. Between the races, physiological differences are not an issue, but between the sexes they are. All the talk of how modern warfare is hi-tech and push-button is off the mark. Ground combat in any setting involves the most physically demanding endurance imaginable. Even in the modern wars, such as those in Afghanistan and Iraq, large numbers of

Table 2.7 Race/gender contrasts in the military

Race	Gender
• Same physical standards	• Different physical standards
• No privacy between races	• Privacy between sexes
• Can advance with racists	• Need enlightened males
• Zero tolerance for discrimination appropriate	• Zero tolerance for harassment excessive
• Officer-enlisted distinction among blacks less salient	• Officer-enlisted among women more salient
• No white anxiety to mentor blacks	• Male anxiety to mentor women
• Emphasize black opportunity	• Emphasize male attitudinal change
• All ranks vulnerable	• Enlisted ranks especially vulnerable
• Need for racially integrated Inspector General (IG)	• Need for female component in Inspector General (IG)

men are involved in physically grueling armored assaults. And, not to be overlooked, much of the work involved in logistics often requires sheer muscle power as well.

Efforts to hold women to the same physical standards as men are delusionary. Rather than trying to raise female standards to abnormal levels, or to lower standards for men, much better to admit the differences and be done with it. It is worth noting that surveys show that women soldiers are quite realistic on this score: 84 percent do not support having the same physical standards for men and women.[21]

The question of personal modesty points to another fundamental difference between race and gender. Whereas privacy within same-sex groups is a non-issue, some level of privacy between the sexes is a primary concern for virtually all military women (and many men too). Nonchalant mixed-sex shower scenes in movies like *G.I. Jane* and *Starship Troopers* (both 1997) to the contrary, both women and men prefer living apart during missions such as those the military have undertaken in the past two decades.

The military can be ruthless on racial discrimination, but 'zero tolerance' for sexual harassment is a non-starter as there is no consensus – in either sex – on what constitutes petty harassment. One person's compliment may be another person's harassment. Likewise, whites do not fear mentoring blacks or vice versa, but a mentor relationship across the sexes can easily lead to innuendo and perceptions of sexual misconduct. This is because the chemistry of sexuality that operates between the sexes has no counterpart in relationships between heterosexuals of the same sex.

Other important differences between race and gender can be pointed out. Blacks can advance within a military system that offers upgrading training and education. Women need more support from enlightened males and do not benefit as a gender per se from concentrated educational programs. Racial discrimination can affect all ranks, while enlisted women are especially vulnerable to sexual harassment. Also, as mentioned earlier, a female chain of complaint may help alleviate sexual harassment, while a racially integrated complaint system works best for racial discrimination cases, as proven in the military.

One other significant difference between race and gender integration must be mentioned. For blacks, the civil rights agenda is the same for both officers and enlisted personnel – equal opportunity for all ranks. There is not a comparable identity of views between female enlisted and female officers in the military. The bottom line is that blacks and whites are essentially interchangeable soldiers. But when physical differences and privacy concerns matter – and they do – men and women are not.

Yet when all is said and done, even the staunchest traditionalist must admit that women bring special talents to the Army. As reported by a presidential commission, women soldiers tend to have higher aptitude scores, better work attitudes and fewer disciplinary problems than the men.[22]

The presence of women soldiers has also been an important – if yet unrecognized – factor in the Army's exemplary performance in recent peacekeeping missions. It is now a matter of record that the behavior of American soldiers toward the local populace in Somalia was superior to that of other armies, including Western ones (not that the behavior of American soldiers was perfect). This welcome outcome was in no small part due to the fact that the American contingent in Somalia was the only one mixed in both race and gender. The forces of all the other contributing nations were all male and all one race (whether white, brown or black).

Conclusion

One other avenue must be explored with regard to race and gender. Why does racial integration work relatively well in the Army, but not so well in the rest of civilian society, notably the university campus? Why does gender integration work relatively well on the campus, but confront major difficulties in the armed forces? This is a topic that deserves greater examination than can be brought to bear here, but one hypothesis can be suggested.

It may be that the intrinsic qualities of a successful military organization and a successful civilian organization have differential meaning for race and gender integration. Jane Jacobs has posited two basic types of authority systems.[23] One, the 'guardianship' model, has the qualities – vertical structure, emphasis on cohesion, loyalty as a core value – that typically characterize a military organization. The 'commercial' model – horizontal structure, emphasis on individualism, honesty as a core value – more closely approximates that of the civilian world. The Jacobs dichotomy parallels one I have made between the armed forces as either an institution or an occupation.[24] An institution stresses cohort identity and group cohesion. An occupation stresses specialized skills and individual merit. It is plausible that an institutional entity operates better as a single-sex entity, while an occupation does so with mixed genders.

Institutional and occupational cultures both operate within contemporary military organizations. But for purposes of generalization, it is reasonable to categorize traditional military organizations as predominantly institutional with guardianship values, and contemporary civilian organizations as predominantly occupational with commercial values. Could it be that the former makes for better race relations and the latter for better gender relations? The implications of this hypothesis are troubling. If it is correct, we cannot expect to have truly trouble-free race relations and gender relations in the same organization.

Notes

The author gratefully acknowledges the support of the Army Research Institute for the Behavioral and Social Sciences (ARI). The mode and presentation of the data are his own sole responsibility and do not necessarily reflect the view of ARI or the US military.

1 The standard reference is B.C. Nalty, *Strength for the Fight: A History of Black Americans in the Military*, New York: Free Press, 1986.
2 S.A. Star, R.M. Williams and S.A. Stouffer, 'Negro Soldiers', in S.A. Stouffer *et al.*,

The American Soldier: Adjustment during Army Life, Princeton, NJ: Princeton University Press, 1949, pp. 486–489.
3 A September 2005 poll reported that 80 percent of blacks compared to 42 percent of whites were not proud of what the United States was doing in Iraq. *New York Times*, 17 September 2005, p. A6.
4 See the website www.dior/whs.mil/mmid/casualty/castop.htm.
5 For a more detailed treatment of race relations in the contemporary American military, see C. Moskos and J.S. Butler, *All That We Can Be: Black Leadership and Racial Integration the Army Way*, New York: Basic Books, 1996.
6 *New York Times*, 24 September 2005, p. A14.
7 On standardized intelligence and academic entrance tests, blacks and Hispanics, on average, score about 20 percentile points lower than do whites or Asian Americans. The Army educational programs described herein greatly reduces, though do not eliminate entirely, the differences between the races. An excellent summary of the issue is given by C. Jencks and M. Phillips (eds), *The Black–White Test Score Gap*, Washington, DC: Brookings Institution, 1998.
8 For the literature on women in the American armed forces, see especially S. Carson Stanley, *Women in the Military*, New York: Mesner, 1993; L. Bird Francke, *Ground Zero: The Gender Wars in the Military*, New York: Simon & Schuster, 1997; B. Mitchell, *Women in the Military*, Washington, DC: Regenery, 1998.
9 Among the more notable reports were those issued by *The Presidential Commission on the Assignment of Women in the Military*, 1993; *The Army Senior Review Panel on Sexual Harassment*, 1997; *The Federal Advisory Committee on Gender-Integrated Training and Related Issues*, 1998; *The Congressional Commission on Military Training and Gender-Related Issues*, 1999. All Washington, DC: US Government Printing Office.
10 C. Moskos, *Enlistment Propensities of University Students*, Washington, DC: Stormingmedia, 2004.
11 W.H. McMichael, *The Mother of All Hooks: The Story of the U.S. Navy's Tailhook Scandal*, New Brunswick, NJ: Transaction, 1997.
12 L.L. Miller, 'Feminism and the Exclusion of Army Women from Combat', *Gender Issues*, Summer 1998, 33–64.
13 Center for Military Readiness, *CMR Notes*, June 2005.
14 Miller, 'Feminism'.
15 See M. Morris, 'By Force of Arms: Rape, War and Military Culture,' *Duke Law Journal*, 45, 4, 1996, 651–781.
16 The time and total number surveyed by gender were as follows: Bosnia, March 1998, male 261, female 42; Kosovo, September 2000, male 260, female 60; Iraq, December 2004, male 430, female 59.
17 Department of Defense, Inspector General Report, *Sexual Harassment and Assault Survey*, March 2005.
18 *Congressional Commission on Military Training and Gender-Related Issues*, op. cit.
19 On women in the Israeli military, see R. Gal, *A Portrait of the Israeli Soldier*, Westport, CT: Greenwood, 1986, pp. 46–58. Also see in the present book Chapter 9 by Edna Lomsky-Feder and Eyal Ben-Ari.
20 *Army Times*, 4 August 1997, p. 31.
21 Reported in L.L. Miller, 'Gender Détente', unpublished doctoral dissertation, Department of Sociology, Northwestern University, Evanston, IL, 1995.
22 *The Presidential Commission on the Assignment of Women in the Armed Forces*, op. cit.
23 J. Jacobs, *Systems of Survival*, New York: Random House, 1991.
24 C. Moskos, 'Institutional and Occupational Trends in Armed Forces', in C.C. Moskos and F.R. Wood (eds), *The Military: More than Just a Job*, McLean, VA: Pergamon-Brassey's, 1998, pp. 15–26. See also C. Moskos, J.A. Williams and D.R. Segal (eds), *The Postmodern Military: Armed Forces after the Cold War*, New York: Oxford University Press, 1999.

3 Diversity in the Canadian forces

Donna Winslow, Phyllis Browne and Angela Febbraro

> The military will launch a vigorous campaign to recruit more visible minorities to make the Forces more reflective of Canada's ethnic diversity.... The image of Canada is that its military projects abroad must be as ethnically diverse as the country back home.... Our population has to look at us and see themselves in us.
> (General R. Hillier, Chief of the Defence Staff, *Ottawa Citizen*, 15 April 2005)

Introduction

In 1971 the Prime Minister of Canada announced a policy of multiculturalism, designed to achieve harmonious intercultural relations by promoting simultaneously cultural maintenance and intergroup contact and participation in the larger Canadian society. As one of the first organizations to become fully bilingual, the Canadian Forces (CF) has become more inclusive in its efforts to reflect the multicultural identity of Canada. Between 1989 and 2001 all barriers to homosexual and female participation in all job qualifications, including combat, were removed. In this chapter we will examine systemic and attitudinal barriers to full participation in the CF by visible minorities, women, and gays and lesbians,[1] as well as the legislation and policies that addressed these barriers.

In the next section of this chapter we will examine policy statements concerning diversity and our theoretical model. We will use the institutional/occupational model, particularly the convergence/divergence approach, to argue that the land element of the CF is the most divergent from civilian society and the most resistant to change. In the third section we will show that legal standards imposed by civilian society lifted systemic barriers to integration and incorporated visible minorities, gays and lesbians, and women into the military. However, even though these groups have equal access to all occupations under Canadian law, there are still attitudinal barriers to their full participation. In the fourth section we will examine these attitudinal barriers and argue that they are more present in the land element of the CF than the other two services – air and sea.

This chapter is based primarily upon documentary research and surveys carried out by the CF on diversity in the Canadian Forces. We also examine how scholars have addressed change within military organizations, and, in particular, how certain sectors of the military react differently to change. This was pertinent since we are particularly interested in how externally imposed legislation has led to change within the CF, and the resistance to that change.

Why is diversity important to the CF?

In this section we will see that the CF ties diversity to military effectiveness. It is argued that military members who work in an atmosphere free of dissension and harassment, one that promotes a strong sense of equal opportunity, are apt to be more productive and team oriented. Any threat to a unit's morale, such as discrimination, is also a threat to its operational effectiveness. Furthermore, according to CF policy it is important that the CF be integral to the society it serves, not isolated from it, and therefore the composition of the military must reflect the population they serve. For example, the Army contends that 'An effective Army that represents more of Canada will inspire the confidence of all Canadians.' Promoting diversity also means an expanded recruitment base and a 'skill enhanced organization more connected to all segments of Canadian society'.[2] Consequently, the current Chief of Defence Staff, General R. Hillier, plans to expand the Canadian Forces by 5,000 full-time and 3,000 part-time reserve personnel and to bolster the ranks of visible minorities. He states, 'We're going to start tilling the ground . . . to go into those ethnic communities across Canada . . . [and seek the] percentages required in order to increase the representation of ethnic minority groups in the CF.'

These statements are interesting in terms of the civilianization debate developed in the post World War II era. Sociologists such as Morris Janowitz observed change in the US military due to technological advances. Changing technology, according to Janowitz, created new patterns of combat and modified organizational behavior in the military. The more complex the technology of warfare, the narrower the differences between military and non-military establishments.[3] Extending this argument, Charles Moskos made reference to a continuum ranging from a military organization highly differentiated from civilian society to one highly convergent with civilian structures. This is popularly known as the institutional/occupational model.[4]

Tension and interplay between institutional (divergence from civilian society) and occupational (convergence with civilian society) tendencies characterize organizational change within the armed forces. Moskos, however, points out that to argue that the military is an institution or an occupation 'is to do an injustice to reality. Both elements have been, and always will be, present in the military system'.[5] This seems to be reflected in what Moskos calls the segmented or plural military. The plural military will be both convergent and divergent with civilian society; it will simultaneously display organizational trends that are civilianized and traditional. The segmented military, Moskos argues, 'will not be an alloy of opposing trends, but a compartmentalization of these trends'. Accordingly, the plural model does not foresee a 'homogeneous military' lying somewhere between the civilianized and traditional poles, but instead, 'the emergent military will be internally segmented into areas which will be either more convergent or more divergent than the present organization of the armed forces'.[6]

Thus, we will see that although the leadership of the CF is supporting convergence with Canadian society, resistance to these policies comes particularly from the combat arms, the most divergent sector of the CF from civilian society. Moskos argues that divergent/traditional features in the military will become most pronounced in labor-intensive support units and combat forces. This sector will 'stress customary modes of military organization', while at the same time 'there will be a convergent sector which operates on principles common to civil administration and corporate structures'.[7] This implies that different services in the military may have different attitudes towards diversity and change.

Legislative and policy framework for diversity in the CF

This section provides a brief overview of legal and constitutional frameworks which have imposed changes upon the CF. The *Report of the Royal Commission on the Status of Women* (1970) marked a turning point for women in the CF. Of the 167 recommendations made to provide a climate of equal opportunity for women in Canada, six pertain to women in the CF.[8] These are: that all trades in the CF be open to women (other than in the primary combat role, at some remote locations, and at sea); the prohibition on the enlistment of married women in the CF be eliminated; the length of the initial engagement be the same for women and men; release from the CF because of pregnancy be prohibited; the CF Superannuation Act be amended so that its provisions would be the same for men and women; and finally, that women be admitted to the Royal Military College of Canada.

Other driving instruments behind diversity in the CF include the Canadian Human Rights Act 1978, which prohibits discrimination on the grounds of race, national or ethnic origin, color, religion, age, sex (including pregnancy and childbirth), marital status, family status, a pardoned conviction, and physical or mental disability (including disfigurement and dependence on drugs or alcohol).[9] Administered by the Canadian Human Rights Commission, which operates independently of the government, the Act has powers of investigation, adjudication and enforcement. This is done through the creation of tribunals that have the power to enforce compliance. So, for example, restrictions on the employment of servicewomen were identified as discriminatory, and the CF had the burden of justifying why women could not occupy certain positions. Despite a concerted effort to muster professional opinion that the restrictions should remain, the CF was unable to convince the tribunal. In 1989 the CF was ordered to integrate women into all roles and environments, except for submarines,[10] but this final barrier was removed in 2001.

The Charter of Rights and Freedoms is part of the Canadian Constitution Act 1981. Section 15, the equality section of the Charter of Rights and Freedoms, came into effect on 17 April 1985, placing more external pressure on the CF. Going beyond the Canadian Human Rights Act, it is a constitutional guarantee of equality – 'the right to the equal protection and equal benefit of the law without discrimination'. It also ensures that discrimination would no longer be tolerated unless bona fide occupational reasons could be cited for the restriction. In its 1985 review of the equal rights provision of the Charter, the Canadian Justice Department cited the CF as potentially in violation of the Charter in six areas: (a) mandatory retirement ages; (b) physical and medical employment; (c) standards; d) recognition of common law relationships; (e) the employment of women; and (f) discrimination based on sexual orientation.

In its 1985 report the Parliamentary Committee on Equality Rights[11] recommended that all trades and occupations in the Canadian Armed Forces be open to women. The committee believed that excluding women from so many job opportunities, most of them indirectly related to combat, had adverse consequences for their careers. It closed to women many well-paid jobs after military service because military training was not available to them; it hindered their promotion in the Forces because they lacked experience in occupations and units that were combat-linked; and it excluded them from experience and training in leadership.[12] The Committee at that time concluded that 'the Canadian Armed Forces must revise its present policy'.[13]

In February 1989 the Canadian Human Rights Tribunal found that the CF discriminated against women by refusing them entry to combat and combat-related employment. This decision removed restrictions on women in combat positions and ordered the complete integration of women within ten years. The tribunal also ruled that the minimum-male requirement be eliminated; that gender-free selection standards be developed; that the integration of women take place with all due speed in order to achieve complete integration in both the reserve and regular forces within ten years; and that this process be subject to external and internal monitoring.

In addition, the Federal Court of Canada ruled that the CF policy of restricting homosexuals from serving was contrary to the Charter of Rights and Freedoms. In October 1992 the CF removed restrictions on homosexuals participating in the military, and in December 1996 same-sex benefits (such as insurance, medical coverage, etc.) were extended to members of the CF.

The Employment Equity Act 1996 (Bill C-64) determined that every Canadian citizen has the right to discrimination-free employment and promotion and that public institutions will strive to be representative of the public they serve. This Act designated four groups for particular attention: (a) women; (b) aboriginal people; (c) persons with disabilities; and (d) visible (racial) minorities. The October 1996 revision to the Act created a legislative framework that included the Royal Canadian Mounted Police, the federal public service and the CF, and it placed certain responsibilities on employers to take steps to ensure that they have an equitable and harassment-free workplace. However, the CF was not legally bound to the provisions of employment equity legislation until November 2002. Owing to the nature of its work, disabled persons were not considered a target group for the CF (other than civilian employees) on account of a request made, and exemption received, in relation to equal disability employment.

Diverse groups in the CF

This section briefly examines the experience of the designated minorities in the three service environments of the CF. It will also discuss the experience of gays and lesbians, though they are not an official minority under the terms of the Employment Equity Act of 1996. Table 3.1 indicates that the CF reflects Canadian society only in terms of linguistic diversity. This results from the requirements of the Official Languages Act. To achieve this, the CF established a quota system in order to have 27 percent of the CF French speaking. However, quotas have since been prohibited under employment equity legislation.[14]

In 2001 the CF administered the Self-Identification Census in order to estimate the representation of designated groups in the CF.[15] This survey had a 73 percent return rate, with 87 percent of respondents self-identifying as a member of a designated group. Of these, 16 percent were women, 3.4 percent were aboriginal peoples, 4.2 percent were visible minorities and 3.1 percent were persons with disabilities. Data also indicated that women represented 10.7 percent of non-commissioned members (NCMs) and 23.3 percent of officers; aboriginal peoples represented 3.3 percent of NCMs and 1.6 percent of officers; and visible minorities represented 3.6 percent of NCMs and 4.3 percent of officers.

Figures released by the Director Military Gender Integration and Employment Equity (DMGIEE) on 1 January 2003 indicated that women accounted for 14.1 percent and 11.9 percent of Canadian officers and NCMs respectively; altogether, 12.4 percent of CF members were women. By late 2003 women comprised 16.3 percent of the CF, including the reserve force, which has had greater success at attracting women than has the regular force. Thus, while women in the reserves are approaching 23 percent, their representation in the regular force appears to have plateaued at around 12 percent.[16] Similarly, data reported by the Department of National Defence in 2004 indicate that the overall representation of women is 12.5 percent in the regular force, 23.8 percent in the reserve force and 7.8 percent among senior officers.[17]

Table 3.1 2001 Census data and demographic trends for designated groups in the Canadian armed forces

Diverse group	% in Canada	% in the CF
Women	50.9	14.9
Aboriginals	3.3	1.5
Visible minorities	13.4	2.6
Persons with disabilities	5.1[a]	1.2

Note
a Includes only persons 15–64 years old but excludes persons institutionalized, on Indian reserves, and in the Yukon, Northwest Territories and Nunavut.

In 2003 about 4.2 percent of the CF regular force members identified themselves as visible minorities,[18] which, in spite of the twofold increase in five years, still falls short of the 9 percent target. The fastest-growing component of the Canadian population, aboriginal peoples constitute only 2.3 percent of the CF regular force. Further, the increase in aboriginal peoples in the CF has been almost exclusively among 'non-status' Indians (that is, Indians who are not formally registered under the Indian Act). Leuprecht observes that the CF attracts more 'non-status' than 'status' Indians, the fastest-growing minority, and concludes that the CF has difficulty recruiting among status Indians, attracting Indians to the officer corps, and retaining Aboriginals.[19] In contrast, aboriginal peoples are over-represented (59.4 percent) in the Canadian Rangers (i.e. part-time reservists who provide a military presence in remote, isolated and coastal Canadian communities).

In light of the situation regarding the designated groups referred to above, it is interesting to note that while the proportion of Francophones in the Canadian population continues to decline, their representation in the CF has changed little in recent years. In 2002, Francophones comprised 22.9 percent of the Canadian population and 27.4 percent of the CF. Thus, the CF has been much more successful in recruiting Francophones than in recruiting designated groups. Success in recruiting visible minorities, for example, may be a function of many complex factors, including internal CF policies and practices, racist societal attitudes, and other sociological and external constraints. For instance, the under-representation of visible minorities may in part reflect the fact that in some countries of origin (e.g. Africa, the Middle East, Asia and South America – increasingly, the countries of origin of immigrants to Canada) the military are seen not as a career of choice or the defender of a nation, but as a machine of oppression.

Visible (racial) minorities

Historically, the three service environments have discriminated against visible minorities. Until 1943 the Navy maintained an explicit racial barrier in official recruitment policy that required personnel to be of 'pure European descent of the white race' before an application would even be accepted. Prior to World War II the Royal Canadian Air Force had maintained a color line stricter than that of the Navy – 'All candidates must be British subjects and of pure European descent. They must also be sons of parents both of whom are (or, if deceased, were at the time of death) British subjects or nationalized British subjects.' In April 1942 the Chief of the General Staff (Army) concluded that 'While Canadian born persons of Japanese origin may appear to be good Canadian citizens, they do, however, bear the appearance and characteristics of another race, which immediately sets them apart from the average Canadian.' The Chief of the General Staff then suggested that any Japanese already in the Canadian Army was to be discharged.[20]

Aboriginal peoples are considered a special minority in Canada because of its colonial history and the efforts made to recognize the special place of First

Nations peoples in the Canadian mosaic. In spite of the large number of status Indians who served in the two world wars, and the more than fifty decorations that were awarded to Aboriginals in World War I, they were still denied the same benefits as non-Aboriginal veterans. In addition to this, they lost their Indian status, thereby forfeiting considerable social benefits. In spite of their contribution to the war effort, an Army recruiting manual of 1944 contained a special note concerning the 'enlistment of Indians and Half Breeds':

> Care should be taken when accepting applications from or approaching Indians as prospective recruits. Here educational standards are strictly adhered to. Experience has shown that they cannot stand long periods of confinement, discipline and the strenuous physical and nervous demand incidental to modern army routine.[21]

The case of Tommy Prince is perhaps the best evidence available to refute this claim. Tommy Prince is generally remembered as one of Canada's most decorated soldiers and certainly as the most decorated Aboriginal soldier in Canadian history. This national and international recognition came from the heroics he displayed during World War II and the Korean War, and the numerous elite medals and decorations he received for his contributions.

There are still attitudinal barriers to the presence of visible minorities in the CF. In a 1996 survey of Regular Force members, results showed that attitudes towards diversity were less positive than in the wider Canadian population.[22] Attitudes towards specific ethnic groups showed that those of European background were rated relatively more positively than those of non-European background. The authors of the study also found a broad pattern of relationships between attitudes and perceptions and the respondents' gender and official language – males and French Canadians being less tolerant of difference. Of particular interest is the observation that attitudes towards diversity differed across the services, with the Sea and Air Forces being more positive than the Land Forces. It is noteworthy that the Land Forces scored higher than the other services on the scale of authoritarianism,[23] are the least tolerant of diversity and equity,[24] and believe in the separation of minorities.[25]

The orientation to authority and its use in hierarchical social arrangements (i.e. authoritarianism) is of importance because there is an established relationship between authoritarianism and ethnocentrism (the tendency to divide people into 'us' and 'them', followed by their differential evaluation as 'us good' and 'them bad', which leads to a low acceptance of diversity and equity).[26]

The data seem to support the idea that the Land/combat forces are more divergent from the general population in their attitudes towards diversity than are the other services. Another indication of this divergence is the report on the Reserves, which shows that the Reserves are intermediate in attitude between the national (civilian) sample respondents and the Regular Force members. However, even though the Reserve Forces had a greater degree of acceptance of diversity and equity and lower authoritarianism than the Regular Force, differences in the Reserve environment (land versus sea) largely replicate those in the Regular Force: the Sea Force was more accepting of diversity and equity and less authoritarian than the Land Forces.[27]

More recent research indicates that some of these patterns have not changed substantially. The 2003 CROP Army Organizational Culture Survey indicated that Army personnel tend to be traditionalists in regard to gender and minorities. In comparison to Canadian society, Army personnel, particularly males, are less supportive of affirmative action for women and minorities; they tend to support sexual stereotypes and wish to preserve their own cultural traditions. In contrast, female Army personnel generally have a more liberal, cosmopolitan outlook and are more sensitive to gender and minority issues.[28] Acceptance of diversity is weakest in the Quebec area. There the CROP survey found elevated levels of ethnic intolerance and the importance of national superiority, two trends that signify a belief in maintaining traditional distinctions among people (by ethnicity, race, religion, etc.).[29] The 2003 CROP Survey also showed that, compared to Canadian society, Army members are more comfortable with structure and discipline and are more willing to respect and defer to authority.[30]

Women[31]

By the end of World War II, 45,691 women had joined the CF (mainly in services support such as administration, communications, logistics and medical support). From 1946 to 1950 there was almost a complete demobilization of servicewomen. All that remained were nursing sisters, whose numbers dropped from 4,480 in wartime to eighty. CF policy stated that a servicewoman's career would end when she married or became pregnant, and this policy remained in effect with the Army until October 1971.

Segal and Segal Wechsler point out that 'military organizations tend to be microcosms of the societies that host them, and thus the rationalization of society and its civilian institutions should be reflected in the military as well'.[32] If this is true, then the CF should reflect the diversity in Canadian society; however, as we have seen, it does not. Even though in 2001 women made up 46 percent of the workforce in Canada, they accounted for only about 16 percent of the CF.[33] However, there are interservice differences. For example, in the 2001 CF Self-identification Census report, women represented 19 percent of the Air Force, 17.9 percent of the Navy and only 12.6 percent of the Army.[34] In 2003, the Committee on Women in NATO forces indicated that women comprise 16.6 percent of the Air Force (14.8 percent of officers and 17.3 percent of NCMs); 11.5 percent of the Navy (15.9 percent of officers and 10.2 percent of NCMs); and 9.6 percent of the Army (11.8 percent of officers and 9.2 percent of NCMs).[35] Further, the concentration of women still continues to be in the more traditional support areas, including medical and dental, with some progress in the less traditional occupations, particularly the naval operations and maritime engineering occupational groups. Modest progress is seen in the combat arms, where women represent 3.9 percent of officers and 1.4 percent of NCMs.[36]

Differences in attitudes regarding the employment of women among the land, sea and air elements have been noticeable since the late 1970s (in questionnaires administered to CF members and their spouses[37] and in national polls of civilians[38]) and the late 1980s. In its 1989 ruling the Human Rights Tribunal made references to the SWINTER (Service Women in the Non-traditional Environments and Roles) trials and noted:

> The evaluation of the air trial concluded that social integration had occurred in a satisfactory manner in a majority of squadrons. Women had performed their tasks well, had received no preferential treatment, and a majority of servicemen agreed that women should be fully employed in previously all male units. The commanding officers believed that this integration was successful and did not compromise effectiveness because both men and women were held to the same high training standards. In other words, the inclusion of women would not detract from but would sustain the esprit de corps.[39]

The tribunal noted that in July 1987 the Minister of National Defence had announced that women were permitted to be employed in all Air Force units. Restrictions were removed and women could enter training for combat aircrew occupations. Major General Morton testified that air units were satisfied that women could be employed in combat roles in mixed-gender units without compromising operational effectiveness.[40]

While these changes were occurring in the air service environment, the sea and land trials revealed a more problematic result:

> Women were judged to perform jobs competently at sea in a supply vessel but neither there nor in the land trials was there satisfactory social integration, from the point of view of all parties. Women complained of the fishbowl effect. Men asserted that women lacked the necessary physical stamina and combat motivation, and received special attention, i.e., favoritism. The Unit's report strongly suggested that many of the problems could be traced to initial poor selection and training, lack of identification of special skills needed, inadequate job definitions, and poor organizational or management preparation.[41]

Thus, the Canadian military seem to be divided into those who represent what Moskos calls an institutional orientation (divergent from civilian society) and those who have an occupational orientation (convergent with civilian society) concerning the integration of women. These attitudes in relation to attitudes and values in the Canadian Army were confirmed in Cotton's study. Cotton found three different 'latent role types' that divided serving personnel. There were the 'soldiers', who 'expressed a strong institutional orientation'; the 'employees', who 'expressed a strong occupational orientation'; and the 'ambivalents', who 'fell between the two extremes'.[42] According to Cotton, 73 percent of 'soldiers' rejected women in combat arms, while 53 percent of 'ambivalents' and

38 percent of the 'employees' rejected them.[43] Data gathered in 2001 show that female participation continues to be highest in the more traditional types of Military Occupational Category (MOC) groups for women (e.g. medical/dental and support) and lowest in the operational types.[44]

In a position paper prepared for the Canadian Department of National Defence, David Segal addressed many of the issues discussed above.[45] He concluded that 'despite the fact that in the short run, as a new phenomenon, gender integration is frequently met with resistance which may in turn constrain cohesion, there is no indication that gender integration negatively affects the performance of military units'.[46] Segal found that 'while there are undeniably problems when previously all-male military formations are gender integrated, there is no firm scientific evidence that would justify the categorical exclusion of women from combat units on organizational grounds'.[47] In relation to physical differences, he argued that there was no question that women and men were different (on the average) in a wide range of 'physical traits related to military performance'. However, he did note that 'physical conditioning increases the capabilities of women and that while average differences between the genders persist after conditioning, the gender distributions tend to overlap'. Thus, where these abilities are relevant to particular military trades, it is rational to screen for the ability rather than for gender, thereby selecting the most qualified individuals'.[48] Again, this demonstrates that many of the arguments used against women are extremely controversial and that often people's personal perceptions and attitudes can have a large impact on what they believe. In relation to Segal's findings, the Human Rights Tribunal cited his observation that: 'the major basis for the categorical exclusion of women from combat units are cultural values regarding appropriate roles for women and resistance from male military personnel'.[49] The tribunal concluded that

> women are, with training, capable of combat roles. The experience of women in combat in the Second World War bears this out. The decision of the air force bears this out. Performance was not an issue as a result of SWINTER trials. Cohesion and the physical and environmental elements are susceptible to management. Integration policies and practices can be designed and applied. We agree with the report of Dr. Segal that attitude is a major factor in making integration work.... Behaviour can to some extent be mandated, with sanctions and rewards as inducements but attitudinal change may not keep pace, and it is this element that must accompany the implementation of an integration policy. Leadership and commitment to integration are essential at the mid and upper levels of command because it is in the operational units that integration must take place.[50]

In brief, allowing women to enter combat positions challenges the traditional ideology that war is a man's world where women do not belong.[51]

Ultimately it was 'external pressure' that led to changes in the combat status of women in Canada, in particular the 1970 Report of the Royal Commission on the Status of Women, the 1978 Canadian Human Rights Act, Section 15 of the Charter of Rights and Freedoms (1985), and the 1989 Canadian Human Rights Tribunal decision on the status of women in the CF (combat arms). It was these laws, regulations and rulings from civilian sources that eventually opened up combat positions for women, and ultimately the full integration of women in the CF.

To date, many countries have had some experience of mixed-gender military combat units, with little evidence of negative impact on effectiveness, cohesion or readiness.[52] From a historical perspective there has been significant success with the integration of women in the combat arms in Canada.[53] However, from a contemporary perspective, progress has been surprisingly slow, when compared with the integration of women into previously all-male domains in the civilian labor force as a whole. Karen Davis and Brian McKee[54] argue that the real hurdle for women in participating fully in the military today still has little to do with their physical and mental abilities but rather revolves around social and cultural issues characterizing a 'warrior' framework. They suggest that military policy and doctrine are increasingly dominated by the terms 'warrior ethos', 'warrior culture' and 'warrior spirit', and that the combat-focused 'warrior framework' has gained emphasis and legitimacy as a schema for describing the entire military. They see this 'warrior creep' as being unwarranted by current and future military requirements (e.g. changes in technology which obviate the need for 'brute strength') and as inimical to the integration of women, as well as increasing proportions of men, as the values and lifestyles of democratic societies evolve. However, in spite of women serving in the combat arms in Canada, they have not served in so-called assaulter roles in Canada's elite anti-terrorist unit, Joint Task Force (JTF) 2. This pattern is echoed by the 2003 CROP survey and the 2004 Army Culture and Climate Survey, which indicate that operational combat units are the least accepting of women, rating them unacceptable in combat and the integration process as marginally successful. In contrast, women rated gender integration more favorably than men, considered the integration process to be progressing, and were more enthusiastic about women in combat.[55]

Since 1999, research in support of the further integration of women in the CF has assumed a low priority.[56] One reason is their superior numerical representation over other designated groups. Another reason is that changing values in society have led to conclusions that, regardless of gender, individuals are impacted by the same issues, with little recognition of inequitable impact and outcome. This perception renders gender as 'neutral' territory in the CF, but, as Karen Davis has argued, it leaves a void in gender research and obscures gender-related inequities.[57]

Gays and lesbians

Homosexuality has been characterized by the Canadian Navy regulations as 'unnatural offenses', by the Army as 'abnormal sexual tendencies', and by the Air Force as 'gross indecency'. Regulations reflected a strong moral aversion to gays and lesbians. Even in the 1960s, Naval authorities viewed homosexuality as an 'evil' and an 'abomination' to be stamped out, and directed attention 'to the dangers – spiritual, moral and physical – of unnatural practices and of condoning these practices in others'.[58]

Homosexuals were classified as deviants and were considered a threat to good order and discipline. In 1967, sexual relations between same-sex consenting adults became no longer a criminal offense. However, until the mid-1970s the process for 'disposal' of homosexual cases remained virtually unchanged.[59] The 1985 review of the equal rights provision of the Charter of Human Rights and Freedoms, which found the CF as potentially in violation, was responsible for six years of study and debate over the issue of gays and lesbians in the military. Surveys of the CF about the impact of allowing homosexuals to serve were conducted in 1986 and 1991; the results showed that military personnel, especially men, were against the lifting of the ban. Male personnel were negative about all aspects of service with homosexuals. 'Most said they would refuse to share living accommodations with known homosexuals and many, particularly those in combat units, would refuse to work with openly homosexual men.'[60]

In spite of CF resistance to the presence of gays and lesbians, the Canadian Federal Court concluded that there was no compelling evidence for the ban and determined that the CF was in violation of the Charter by restricting the military service of homosexuals. The CF complied with the court's decision and lifted the ban on homosexuals in 1992. According to several authors, the integration of homosexuals into the CF has been a relative success because the approach was deliberately low-key.[61]

So far, the policy changes and their practical implementation seem to have progressed smoothly. None of the dire predictions about performance, recruitment, retention and violent reactions have proved correct. Part of the reason for the successful implementation was the immediate and full support of the policy change by the leaders in the Department of National Defence and the CF. 'They made it a leadership responsibility at every level of the CF and showed they meant to implement the policy fully.'[62] Moreover, different from the issue of gender integration, special attention was directed away from homosexuals rather than towards them as a special case. Still, the 2004 Army Culture and Climate Survey and the 2003 CROP Army Organizational Culture Survey[63] suggest that while diversity is generally well accepted in the Army, a large segment of males indicate that gays and lesbians are not acceptable as workmates, indicating that attitudinal barriers are still prevalent.

Conclusions

We have seen that the CF has been compelled to change by forces external to the organization – particularly laws and legislation that have imposed diversity and gender integration. It has been insufficient, for the most part, for the CF to argue that it is a unique institution and therefore exempt from some of the conditions of the Canadian Human Rights Act or the Charter of Rights and Freedoms. Therefore, even though the CF has largely been resistant to these changes, it has been compelled to comply.

The CF leadership and the organization have now undertaken a large number of initiatives to promote diversity and make the organization more convergent with civilian society. Nevertheless, there is concern about the Canadian military's tendency towards authoritarianism. The 1997 CF survey suggests that 80 percent of the Regular Force sample and 71 percent of the Reservists are on the high side of the authoritarian scales (compared to 50 percent in Canadian university samples).[64] Moreover, authoritarianism is concentrated in the land element. Research in Canada[65] has described authoritarianism as a readiness to submit to a higher authority that is perceived to be legitimate – an attitude which is encouraged by the military. It can also mean aggressiveness, which is described as a tendency to act aggressively or negatively towards those who are different, particularly when sanctioned by established authority; this tendency is in a way also encouraged by the military for the purpose of combat. According to Berry and Kalin,[66] these dispositions are very probably incompatible with the acceptance of diversity and equity. These authors also maintain that there is

> long-standing and contemporary evidence of a link between ethnocentrism and authoritarianism: those who are intolerant of diversity (and reject employment equity principles and practices) tend to be those who espouse authoritarian views that are often present in 'tight' and 'hierarchical' organizations.[67]

The CF diversity survey results, which indicate that the Land element seems to be more authoritarian and less tolerant of diversity and female integration than the other elements, have also been confirmed more recently by the results of the 2004 Army Culture and Climate Survey and the 2003 CROP Army Organizational Culture Survey.

Ethnocentrism involves very basic psychological processes of social differentiation into in-groups and out-groups (us versus them), followed by differential evaluation (us good versus them bad) and differential treatment or discrimination.[68] The difficulty for military forces is that their combat training encourages social differentiation into in-groups and out-groups in order to categorize people as the 'enemy', who then can be killed. How then is it possible to turn this training 'off' in complex environments such as peace operations? The CF's experience in Somalia in the mid-1990s, described by Winslow,[69] indicates that some members of the CF Airborne Regiment perceived the Somalis as different and 'other'. This was then followed by their derogation and rejection, and ultimately contributed to the torture and subsequent death of a Somali teenager. How is it possible to create peace operation attitudes where differences should be accepted and protected rather than attacked? This question will become ever more pertinent in the context of a security environment that is becoming increasingly characterized by multinational and multicultural military operations, and by the 'three-block war' concept, in which military forces will be required to

44 *Winslow* et al.

conduct combat operations, peacekeeping operations and the provision of humanitarian assistance all within a short span of time and space.

If diversity is to be celebrated and promoted in Canadian society and protected by legislation, then how can processes of differentiation encouraged by combat training lead to positive evaluation of the 'other' and to equitable treatment? If militaries are to remain focused upon combat and hierarchy, promoting diversity may be more difficult than anticipated. Yet they must also reflect the values of the larger society if they are to maintain the allegiance and support of those whom they are supposed to serve.

To conclude, evidence to date suggests a mixed picture concerning diversity in the CF. Although progress has been made in some areas, the CF still faces challenges in recruiting women, visible minorities and aboriginal peoples, especially in the Regular Force and in the officer corps.[70] The situation is even more critical when we look at projections for the future. The CF estimates that by 2011 the population comprising people who do not identify as Anglophone, Francophone or Canadian will reach 39 percent, and racial minorities will represent 19.2 percent of the Canadian population (and will increase to 21.1 percent by 2016). Should current recruitment trends persist, the CF's composition will diverge increasingly from the ethno-demographic composition of Canadian society.[71]

Notes

1 For the purpose of this chapter we will consider systemic barriers to diversity as the policies, guidelines or procedures that intentionally or unintentionally favor one group over another, and attitudinal barriers to diversity as the attitudes, beliefs and/or behaviors that can lead to a non-supportive work culture and environment.
2 'Leadership in a Diverse Army: The Challenge, the Promise, the Plan', mimeo/first draft, Ottawa: Canadian Army, 1988, p. 1.
3 M. Janowitz, *The New Military: Changing Patterns of Organization*, New York: Russell Sage Foundation, 1964.
4 C. Moskos, *Soldiers and Sociology*, Washington, DC: United States Army Research Institute for the Behavioral and Social Sciences, 1988, p. 57.
5 Ibid.
6 C. Moskos, 'The Emergent Military: Civil, Traditional or Plural?', *Pacific Sociological Review*, 16, 2, 1973, 275.
7 Ibid., p. 277.
8 K. Davis, *Women and Military Service in Canada: 1885–Present*, Canadian Forces Personnel Applied Research Unit, Historical Report 96-1, 1996, p. 7.
9 F. Pinch, *Perspectives on Organizational Change in the Canadian Forces*, Research Report 1657 (NTIS No. ADA277746), Alexandria, VA: Army Research Institute for the Behavioral and Social Sciences, 1984, p. 7.
10 Ibid., p. 8.
11 The Parliamentary Committee on Equality Rights was established to examine areas that might be in conflict with section 15 of the Charter of Rights and Freedoms.
12 Cited in *Human Rights Tribunal Decision*, TD 3/89 (in the matter of a Hearing between the Complainants and the Canadian Armed Forces), Ottawa, 20 February 1989, p. 29.
13 Ibid., p. 30.

14 Instead, the CF has established long-term (twenty-year) recruitment goals of 9 percent for visible minorities, 28 percent for women, and 3 percent for aboriginal peoples.
15 N.J. Holden, *The Canadian Forces 2001 Self-Identification Census: Methodology and Preliminary Results*, Director Military Gender Integration and Employment Equity and Director Strategic Human Resources, D Strat HR Research Note RN 01/03, 2003.
16 C. Leuprecht, 'Demographics and Diversity Issues in Canadian Military Participation', in F.C. Pinch, A.T. MacIntyre and P. Browne (eds), *Challenge and Change in the Military: Gender and Diversity Issues*, Canadian Forces Leadership Institute, Canadian Defence Academy, Winnipeg, Manitoba: Wing Publishing Office, 17 Wing Winnipeg, 2004.
17 K.D. Davis, 'Research and Organizational Change: Women in the Canadian Forces', *Canadian Journal of Police and Security Services*, 2(3), 2004, 171–180.
18 Leuprecht, 'Demographics and diversity issues'.
19 Ibid.
20 Canadian Forces, *Diversity Training Program Handbook Cpl and Pte*, Ottawa: Department of National Defence, 1998, p. 13.
21 Ibid., p. 14.
22 Multicultural Associates Inc., *Canadian Forces Diversity Project: Baseline Survey*, Kingston, Ontario: Department of Psychology, Queen's University, 1997, p. 2.
23 J.W. Berry and R. Kalin, *Canadian Forces Diversity Project: Baseline Study*, Report to the Canadian Forces Personnel Research Team (CFPRT), Kingston, Ontario: Department of Psychology, Queen's University, Multicultural Associates Inc., 1997, p. 9.
24 Government of Canada, *Employment Equity Principles for the Canadian Forces*, Ottawa, 1993.
25 J.W. Berry and R. Kalin, *Canadian Forces Diversity Project: Baseline Survey*, Report to the CFPRT, Kingston, Ontario: Department of Psychology, Queen's University, Multicultural Associates Inc., 1997, p. 24.
26 J.W. Berry and R. Kalin, *A Conceptual Framework for Achieving Diversity and Equity in the Canadian Forces*, Report to the CFPRT, Kingston, Ontario: Department of Psychology, Queen's University, Multicultural Associates Inc., 1997, p. 2.
27 Ibid.
28 K. Farley and B. Wild, *The Army and Canadian Society: A Socio-cultural Comparison*, Land Personnel Concepts and Policies, National Defence Headquarters, Ottawa, 2005.
29 M. Capstick, K. Farley, W. Wild and M.A. Parkes, *Canada's Soldiers: Military Ethos and Canadian Values in the 21st Century Army*, The Major Findings of the Army Culture and Climate Survey and the Army Socio-Cultural Survey (CROP 3SC). Report to the Commander Land Force Command, 2004.
30 Farley and Wild, *The Army and Canadian Society*.
31 This section draws upon a more extended study of gender integration in the CF. See D. Winslow and J. Dunn, 'Women in the Canadian Forces', in A. Sens (ed.), *Canadian Security and Defence Policy: Strategies and Debates at the Beginning of the 21st Century*, Vancouver: University of British Columbia Press, 1999.
32 D. Segal and M. Segal Wechsler, 'Change in Military Organizations', *Annual Review of Sociology*, 9, 1983, 151.
33 Holden, *The Canadian Forces 2001 Self-Identification Census*.
34 Ibid.
35 Committee on Women in NATO Forces, *Proceedings from the Meeting of the Committee on Women in NATO Forces*, Ottawa, 2003.
36 Ibid.
37 Directorate Personnel Development Studies, *Summary of Report of the Attitude Surveys on Women in Combat Roles and Isolated Postings*, Study Report, June 1978.
38 The polls were conducted by Gallup Canada for the Department of National Defence,

46 *Winslow* et al.

and assessed public attitudes (Canadian public opinion) towards women serving as soldiers, sailors and aircrew.
39 *Human Rights Tribunal Decision*, T.D. 3/89 (in the matter of a hearing between the complainants and the Canadian Armed Forces), Ottawa, 20 February 1989, p. 26.
40 Ibid., p. 35.
41 Ibid., pp. 26–27.
42 C. Cotton and F. Pinch, 'The Winds of Change: Manning the Canadian Enlisted Force', in D. Segal and H. Sinaiko (eds), *Life in the Rank and File*, Washington, DC: Pergamon-Brassey's, 1986, p. 242.
43 C. Cotton, *Military Attitudes and Values of the Army in Canada*, Canadian Forces Personnel Applied Research Unit, Research Report 79-5, 1979, p. 86.
44 Holden, *The Canadian Forces 2001 Self-Identification Census*. See also L. Tanner, *Gender Integration in the Canadian Forces: A Quantitative and Qualitative Analysis*, Director General Military Human Resources Policy and Planning, Directorate of Military Gender Integration and Employment Equity and Director General Operational Research Directorate of Operational Research (Corporate, Air, and Maritime), ORD Report R9901, 1999.
45 Similar arguments are put forth in K.D. Davis and B. McKee, 'Women in the Military: Facing the Warrior Framework', in Pinch *et al.* (eds), *Challenge and Change in the Military*.
46 D. Segal, *The Impact of Gender Integration on the Cohesion, Morale, and Combat Effectiveness of Military Units*, Position Paper for the Canadian Department of National Defence, Annex K to Part 3 of the Charter Task Force Final Report, 1986, p. 19.
47 Ibid., p. 24.
48 Ibid.
49 Ibid., p. 59.
50 *Human Rights Tribunal Decision*, op. cit., p. 60.
51 The inclusion of women in combat is still controversial among some sectors of the Canadian public, as indicated by a letter by newspaper columnist Barbara Kay, 'Fighting Is For Men', *National Post*, 5 November 2005.
52 Davis and McKee, 'Women in the Military'.
53 One sign of success is the fact that many women leaders in the combat arms do not feel that they must 'act like men' in order to be perceived as effective leaders. A.R. Febbraro, 'Gender and Leadership in the Canadian Forces Combat Arms: A Qualitative Study of Assimilation vs. Integration', *Canadian Journal of Police and Security Services*, 2, 4, 2004, 215–228. Similarly, another measure of progress in terms of gender integration can be seen in the decline in the rate of sexual harassment reported by women in the CF, from 26 percent in 1992 to 14 percent in 1998. See N.J. Holden and K.D. Davis, 'Harassment in the Military: Cross-national Comparisons', in Pinch *et al.* (eds), *Challenge and Change in the Military*.
54 Davis and McKee, 'Women in the Military'.
55 Capstick *et al.*, *Canada's Soldiers*.
56 Davis, *Women and Military Service in Canada*.
57 Ibid.
58 Canadian Department of National Defence, 1962, quoted in Pinch, *Perspectives on Organizational Change*, p. 11.
59 Ibid., p. 12.
60 P. Gade, D. Segal and E. Johnson, 'The Experience of Foreign Militaries', in G. Herek, J. Jobe and R. Carney (eds), *Out in Force: Sexual Orientation and the Military*, Chicago: University of Chicago Press, 1996, pp. 106–130.
61 Ibid.
62 Ibid.
63 Capstick *et al.*, *Canada's Soldiers*.
64 Berry and Kalin, *Conceptual Framework for Achieving Diversity and Equity*, p. 32.

65 Altemeyer, 1988, cited in Berry and Kalin, *Conceptual Framework*, p. 32.
66 Berry and Kalin, *Conceptual Framework*, p. 26.
67 Ibid.
68 M. Patchen, *Diversity and Unity: Relations between Racial and Ethnic Groups*, Chicago: Nelson-Hall, 1999, pp. 32–35.
69 D. Winslow, *The Canadian Airborne Regiment in Somalia: A Socio-cultural Inquiry*, Ottawa: Minister of Public Works and Government Services Canada, 1997.
70 Leuprecht, 'Demographics and diversity issues'.
71 Ibid.

4 Indigenous integration into the Bolivian and Ecuadorean armed forces

Brian R. Selmeski

Introduction

Effectively integrating large but previously marginalized indigenous populations to state institutions is one of the greatest challenges to consolidating democratic governance in the Andean region. This is particularly true in the highly diverse central republics of Bolivia, Peru and Ecuador. Integration is significantly complicated by native people's and organizations' legitimate desires to retain meaningful elements of their cultures and identities while acquiring the material and symbolic benefits of first-class citizenship. Dominant groups, on the other hand, often assume that *indígenas* (indigenes) will – and, more importantly, should – become more like them as a result of the process. Previous state policies premised on acculturation or assimilation are simply incapable of reconciling the competing expectations of these groups. Tensions are particularly salient and vexing in the armed forces, given their traditional emphasis on cultural homogeneity and important role in the region's socio-political milieu. This makes their study both intellectually alluring for scholars and significant for all those involved in the process.

This chapter examines how two of the region's armies have grappled with these issues. The data are drawn from the author's doctoral fieldwork in Ecuador (1999–2002) and applied research in Bolivia (2003–06). After providing a brief background of the militaries and their respective societies, the focus will turn to indigenous–military relations in the contemporary era – roughly since 1980, with particular emphasis on the post-1990 years. By examining similarities and differences between the cases on the empirical, theoretical and contextual levels, I seek to make this chapter useful to a wider spectrum of both academics and practitioners. The conclusion continues this approach by synthesizing some of the key lessons and strengths from the cases, hopefully providing a modicum of insight to those seeking to promote military diversity in the Andes and elsewhere.

On the surface, both cases entail issues of the armed forces' representation and openness that have great significance for recruitment, operational effectiveness and civil–military relations in their respective countries. Underlying these concerns, however, are a deeper set of issues, particularly the nature of citizenship and national identity in a diverse society. The Ecuadorean and Bolivian militaries have therefore become key – but not the sole – sites for debating and defining broader state issues. In part, this reflects the relative strength and importance of the region's armies and indigenous organizations, yet it is also indicative of the nature of the academic and practical debate.

As Dietz and Elkin noted, any discussion of integration must, by its very nature, address the military, the rest of the state, and society, from which both institutions emanate.[1] Restricting the focus to internal military integration alone would artificially truncate inquiry and introduce a series of false assumptions to the analysis. Even the most self-regulating armed forces are not completely autonomous from governmental control or immune to societal pressures. Thus, the approach taken here is to contextualize the experiences of native soldiers within the larger questions of indigenous–state relations, national identity, and the role of the military in governance as well as society.

Moreover, as the Bolivian and Ecuadorean cases demonstrate, successful integration of national minorities to the armed forces entails far more than just the presence of diverse individuals in the ranks. It also requires a shared sense of identity and commitment to the core values that inform such a project, including equity, respect and diversity. Whether these emanate from part or all of society, are imposed by government, or are developed unilaterally by the armed forces, such notions have significant implications for all three groups. Consequently, in this chapter I apply a limited interpretation of integration, to reflect not just the presence of national minorities in the armed forces, but also their just treatment.

Fundamentally, this includes institutional recognition of their legitimate desire to maintain many – though not necessarily all – of their differences. By this measure, the inclusion of indigenous peoples in the military is not integrative if their differences are erased or minimized in the process. Thus, while quantitative measures are necessary to assess the scope of indigenous presence in the armed forces, qualitative factors permit us to gauge both whether these efforts qualify as integration and their nature. To this end, I differentiate between accommodation, which promotes the incorporation of individuals from distinct cultural backgrounds through tolerance, and acceptance, which truly values differences as a core institutional principle.

Accommodation policies can range from the removal of employment barriers (physical, social, cultural and other) to the establishment of quotas for designated groups. Such policies promote greater presence and awareness of differences without generating the concomitant understanding and respect between groups. Accommodation efforts are usually limited, to ensure they do not become onerous or threaten closely held military beliefs and traditions. I have therefore dubbed the results 'many-culturalism' to distinguish them from bona fide 'multiculturalism', which I argue, can only be achieved through the establishment of a culture of acceptance.[2]

Whereas multiculturalism tends to rest on ideological rationales (i.e. consistency with the principles of liberal democracy), accommodation usually employs practical ones (legal compliance, military benefit, etc.). In practice, however, the two approaches are often difficult to distinguish. In the best of circumstances they may reinforce each other, but, as the cases explored here demonstrate, accommodation may also unintentionally limit efforts to achieve acceptance by serving as an escape valve for social pressures.

Historical context

Contemporary efforts to integrate indigenous peoples to the Bolivian and Ecuadorean armies in these manners are relatively new, and break with decades of alternating efforts to assimilate Indians into the military, or exclude them altogether. The early leaders of Ecuador and Bolivia were faced with large territories and weak state institutions after independence in the early 1800s. Consequently, they frequently relied on force to rule. Some elements of the armies of independence were cobbled into national militias, but other groups devolved into armed bands led by local strongmen.[3] Whereas politicians generally appointed officers from the white elites, raising armies from an ethnically diverse population proved to be more difficult. To do so, political and military leaders had to reconcile competing and at times mutually exclusive requirements.

On the one hand, leaders had to create an armed force that was capable of defending the country from foreign and domestic threats throughout the entire national territory. This required raising sizeable forces, with poorer and purportedly inferior men logically assigned to fill the ranks. In raising their forces, political and military elites had to decide whether *indígenas* should serve at all. At first the answer was often 'no', but when the response was affirmative, elites had to assign Indians martial roles that reflected the racial assumptions of the national vision. Finally, leaders sought to craft a military that was subordinate to civil (though not always civilian) authority and reliable – willing to follow orders and not turn on leaders through coups or rebellions. Groups that were deemed trustworthy were recruited, whereas those believed to pose a potential threat were frequently excluded or marginalized.

These tensions generated significant consequences for indigenous–military relations in general and the recruitment/treatment of Indian soldiers in particular. Andean armies have at times been required to sacrifice ideological beliefs about Indians' appropriate military roles to practical 'manpower' considerations in order to ensure the collective defence. On other occasions, deeply held prejudices have impinged on troop strength by excluding *indígenas* from the pool of eligible soldiers. Likewise, concerns that training Indians to fight in the Army would threaten rather than enhance state security have occasionally limited their recruitment.

Finally, when conscripted, Indians have not always been afforded equal or fair treatment, frequently being subjected to systematic acculturation efforts as part of the national project. Thus, in addition to providing the state with a mechanism to mobilize soldiers, conscription has also served as a system of moral regulation. This coupled racial and national discourses with those of citizenship, often producing terrible conditions of service for indigenous recruits.

These issues came to the fore with the first wave of military professionalization in the late 1800s and early 1900s.

Prior to this period, *indígenas* had either been legally exempted from military service, protected by large landholders keen to maintain control of their labour pool, or ignored by prejudiced recruiting officials in Ecuador. This changed with the Liberal Revolution of 1895, which emphasized respect for individual liberties, particularly those of Indians, who were portrayed as equal citizens in the eyes of the state. President Eloy Alfaro quickly established a permanent standing army and contracted a Prussian-trained Chilean military mission in 1900 to guide this effort. One of the Chileans' first accomplishments was to draft a European-style conscription law, which, ironically, exempted Indians from service.[4] This provision was revoked in 1920, and by the 1930s, conscription was gradually implemented in the highly indigenous central provinces. Yet during the same period an Italian military mission introduced eugenic criteria for admission to the Military College.[5] The result was to institutionalize the traditional division of labour, effectively banning Indians from the officer corps while obliging them to provide the bulk of conscripts.

These changes were heavily influenced by *indigenismo*, a paternalistic – though, at the time, progressive – discourse of political elites to protect Indians and incorporate them into the nation. Fuelled by the Mexican Revolution of 1910 and events in neighbouring Peru, *indigenismo* sought to erase Indians' distinctiveness either by subordinating cultural considerations to economic ones or by actively promoting *mestizaje* (cultural mixing). Gnerre and Bottasso distinguish between 'state-sponsored *indigenismo*' and 'supra-national' forms that intellectuals favoured.[6] Both approaches were present in early twentieth-century Ecuador, with academics and officers concurring that conscription was one of the primary vehicles for accomplishing their objectives.[7] Implementing these ideals, however, proved to be far more difficult in Ecuador than Bolivia.

Bolivian Indians too had largely been excluded from the post-independence military. In addition to racial prejudices, both countries' leaders were concerned with the potential security threat of militarizing the indigenous majority. Economic interests were also influential: Bolivian mine owners were eager to maintain control over Indian labourers, as were Ecuadorean agricultural elites. Yet it took the disastrous War of the Pacific (1879–80), in which Bolivia lost its nitrate-rich littoral to Chile, to reveal both the incompetence of the country's generals and the weakness of its Army. As in Ecuador, the government subsequently contracted foreign military missions – first Prussian, then French – to help professionalize the armed forces, paying particular attention to obligatory military service.

In addition to raising a mass army, the conscription law sought to control the indigenous population physically and morally. The importance of such measures was reinforced during the Federal War (1898–99), when numerous Indians rallied behind Aymara leader Zaraté Willka to defend their lands from encroachment by mining and agricultural elites. The intent of conscription in Bolivia was therefore similar to that in Ecuador. In the words of the country's leading military sociologist, it sought to resolve the indigenous question by their 'civilization' and at the same time [establish] 'citizen equality' regarding national defence. Egalitarian rhetoric and the formalization of the principle of universality notwithstanding, the new law did nothing more than to legitimate exclusionary and discriminatory practices.[8]

The government soon established complex administrative procedures and financial expenses favouring the urban middle class, and *indígenas* quickly came to fill the ranks of the Bolivian Army.

As in Ecuador, Bolivian indigenous conscripts were simultaneously valued for their ability to endure physical hardship and denigrated for their supposedly inferior culture. Likewise, Indians were systematically excluded from the Military College, a prerequisite for officership in the newly professionalized Army: from its establishment in 1912 until 1952, applicants had to demonstrate proof of legitimacy (i.e. that they were the baptized product of a sanctioned marriage) and honourable status (i.e. that they were from a socially and economically well-off family).

These inequities were laid bare in both countries during the mid-twentieth century as a result of failed military campaigns and significant territorial losses. However, the militaries' distinct responses to these challenges heralded future policy differences. Ecuador's disastrous 1941 border war with Peru revealed the military's inability to mobilize ex-conscripts in the region as well as gross mismanagement by generals, poor infrastructure and antiquated armament. Peru occupied much of southern Ecuador, which quickly sued for peace and forfeited an enormous portion of the Amazon it had claimed – nearly half its national territory. This, together with the war in Europe, ushered in a new era of militarization of Ecuadorean society. The following year President Arroyo del Río toughened the conscription law by decree, and a year later Congress declared pre-military instruction obligatory for all students over 12 years of age.

Mercedes Prieto observed that in the aftermath of defeat, 'elites timidly discussed Indian participation in the army' and noted that they 'had exhibited "una clarinetada de patriotismo" (a bugle call to patriotism)'.[9] Yet while these laws greatly increased public participation in the armed forces, Indians remained largely outside the military apparatus. This was in part a result of elites' concern that if *indígenas* were conscripted in great numbers – and de-Indianized in the process – they might not return to the countryside upon completion of their service. In 1944, President Velasco Ibarra brokered a compromise to this dilemma with the establishment of two indigenous work battalions under the *conscripción indígena rural* (rural indigenous conscription, CIR) program.[10] This way, Indians could contribute to national defence without threatening the established social order.

In Bolivia, on the other hand, forced conscription of *indígenas* increased dramatically during the disastrous Chaco War with Paraguay. Of the roughly 250,000 men mobilized for combat, over one-quarter never returned, many falling victim to disease and dehydration.[11] Indians bore the brunt of the suffering, although in the aftermath of defeat, white and *mestizo* veterans – the 'Chaco generation' – also began to question the military and national institutions in general. The Army largely ignored the public discontent, and continued drafting large numbers of *indígenas* to fill the ranks.

This came to an abrupt halt in 1952, when miners and peasants of the Nationalist Revolutionary Movement (MNR) vanquished the Army in three days of clashes. The following year the MNR reduced conscription by half, temporarily emasculating the military in favour of armed militias that fused class and ethnic interests. Simultaneously, the Military College was shuttered and hundreds of officers purged from the institution. Despite popular perception that the Army had ceased to exist, it gradually regained its pre-revolutionary stature as the government increasingly came to fear the power of the indigenous peasant militias.[12] In the process, old patterns of conscripting Indians and commissioning whites and *mestizos* returned.

Though military service was unpopular, indigenous and peasant leaders tolerated it during the era of dictatorships (1964–82) as a result of the *Pacto Militar–Campesino* (Military–Peasant Pact) and the Army's vigorous civic action programmes.[13] Both were military efforts to prevent the spread of communist insurgency among the indigenous population, a real concern for many in both Washington and La Paz after the failed exploits of Che Guevara in Bolivia during 1966–67. The epicentre for indigenous radicals was the mining communities, where Indians had laboured since conquest, extracting first silver, then tin, for international interests. Regular clashes between the Army and miners left dead and wounded on both sides, but particularly the workers, as well as bitter memories of excessive violence and brutality at the hands of the security forces.[14]

The Ecuadorean Army took a different approach to these Cold War era concerns. After seizing power from President Carlos Julio Arosemena in 1963, a military *junta* ruled the country until 1966. Rather than using sheer force, this politically conservative but economically progressive administration sought to prevent communist inroads by modernizing the country and promoting socio-economic development, starting with agrarian reform. The *junta*'s revisions to the Constitution and other laws also facilitated the creation of *conscripción agraria militar ecuatoriana* (Ecuadorean military agrarian conscription, CAME) in 1967.

In many ways, CAME mirrored CIR, yet twenty-two years had passed since the earlier programme's inception, and its *indigenista* justification was no longer adequately reflected in the military's vision or goals. The official plan acknowledged that CAME would incorporate Indians into 'the active life of the country',[15] but *indígenas* were only one of many groups the programme targeted. Non-indigenous peasants from the highlands and coastal region would also participate and learn technical agricultural methods while acquiring a new, national identity as *mestizo* citizens.

Participants were referred to simply as *campesinos* (peasants) in official documents, effectively erasing Indians – and the 'Indian problem' – from the administration's political agenda. Like the Peruvian administration of General Juan Velasco (1968–75), the Ecuadorean *junta* 'officialized the disappearance of Indians as Indians, [by] recognizing them instead only as peasants'.[16] The military's redefinition of Indians as *campesinos* did not erase flesh and blood *indígenas*, nor their (and others') sense of their difference. The Army was not blind to this, and introduced the first accommodation measures, permitting indigenous men from the Otavalo region to retain their long hair while in uniform.[17] (Indians' persistence came back to haunt the Army decades later, who responded with further accommodations.)

Reclassifying *indígenas* as *campesinos* did, however, reinforce the military's belief that Indians are defined primarily by their mode of production. It also provided a discursive means of anchoring conscription to *mestizaje*: in *mestizaje*, officers saw a means of national unification that fitted their larger objectives, while conscription (and CAME in particular) provided them with an ideal mechanism through which to transform Indians (now peasants) into *mestizos*. The military and other elites came to conceive of the Ecuadorean nation in ethnic terms – as a *mestizo* nation. Theoretically, all inhabitants could become citizens, to a greater or lesser extent, without miscegenation by shedding their 'otherness' (i.e. 'Indianness' or 'blackness') and becoming cultural *mestizos*. This discourses of *mestizaje* drew heavily on *blanqueamiento* (whitening) and postulated that blacks and Indians should, and would if properly motivated, voluntarily abandon their respective cultures in favour of white culture in order to move up the ethnic hierarchy.[18]

Ecuadorean *mestizaje* continued to emphasize culture over genetics and behaviour over skin colour, but it was never ideologically neutral. *Mestizaje* was a highly directional, charged and biased project that was intimately linked to nation building. Whites and *mestizos* were not expected to become more indigenous, but vice versa. This was a result of assumptions 'that ethnicity and nationality tend to mutual exclusivity and that the cultural aspects of being ethnic are not simply a function of the national political economy'.[19] In other words, the military could never achieve national unity through development alone; Indians had to be eliminated by persuading (or forcing) them to adopt racially neutral national identities and values as *mestizos*.

Thus, by the return to democracy in the 1980s, both the Bolivian and the Ecuadorean Armies remained highly stratified, with Indians relegated to the lowest echelons. However, whereas the Ecuadorean officers had created conceptual linkages between the groups and programmes to operationalize these beliefs, their Bolivian counterparts continued to accept and implement a caste-like social system. This key difference, together with external pressures, has largely shaped the nature of indigenous–military relations in the intervening years. The result has been uneasy but generally peaceful accommodation between the Ecuadorean Army and indigenous groups, whereas the Bolivian Army and *indígenas* have experienced increased conflict.

Contemporary multicultural efforts

The Ecuadorean Army's ability to avoid wide-ranging conflict with the country's indigenous movement – arguably the most organized and powerful in the hemisphere – required a major shift in institutional attitudes regarding national identity and the objectives of conscription. The need became clear on 4 June 1990, when the Confederación de Nacionalidades Indígenas del Ecuador (Confederation of Indigenous Nationalities of Ecuador, CONAIE) orchestrated its first national strike. The protest, or *Levantamiento* (Uprising), as it came to be known, shut down the crucial Pan-American Highway, paralysing the country, and occupied symbolic public buildings. This placed the Army in a difficult position: on the one hand, the armed forces are constitutionally charged with maintaining domestic order and were therefore responsible for confronting the civil unrest when it surpassed the National Police's capabilities. On the other hand, there was empathy in the ranks for *indígenas*' plight. This tempered the Army's response[20] and led to prolonged negotiations between the indigenous movement and government.

The tension persisted within the armed forces, though. Initially, they characterized the *Levantamiento* as 'subversive' and the work of foreign agitators seeking to undermine national unity by creating 'a state within a state'.[21] Gradually, CONAIE's words and deeds softened some members' positions. Portions of the indigenous agenda – particularly those framed in a discourse of economic development, land reform and individual liberties – resonated with many officers. These were national problems, one Army officer noted, and therefore it was 'necessary to make whites and *mestizos* aware that the problem of the Indian is everyone's problem'.[22] Yet most of these officers continued to define 'the problems' as *indígenas*' illiteracy, lack of land, discrimination, and the dearth of state services in rural areas.[23]

More fundamentally, the *Levantamiento* directly challenged the state's monopoly on defining the nature of the Ecuadorean nation. *Indígenas* insisted that they not only remained *as indígenas* – despite decades of *mestizaje* – but also were simultaneously Ecuadoreans and members of indigenous nations. This language generated significant concern within the armed forces, which categorically rejected CONAIE's characterization of Ecuador as a pluri-national state. In the wake of the Uprising, military professional journals and educational institutes became increasingly focused on *indígenas* and questions of national identity/unity. The conclusion of these studies and debates was as obvious as it was irrefutable: the *mestizaje* project had failed. The alternative was not, however, immediately clear.

Initially, the military high command issued an order 'to develop cooperation and approximation programmes between the armed forces and the country's indigenous communities'.[24] While ostensibly these programmes were efforts to improve the well-being of *indígenas*, privately officers confessed concern over infiltration of Sendero Luminoso (Shining Path) from Peru. The Army soon institutionalized these efforts by establishing *compañías de acción cívica y forestación* (Civic Affairs and Forestry Companies, CACYF), which still exist.[25] CACYF drew on and modified earlier, Vietnam era-inspired counter-insurgency projects by providing medical, dental and educational services; constructing basic infrastructure; and promoting the development of micro-enterprises in indigenous communities neglected by the state. These policies paid dividends in January 1995, when Ecuador and Peru went to war over a lingering border dispute. Commanders and observers alike were ebullient in their praise for local indigenous groups' support, with some describing the cooperation as a key to Ecuadorean victory.[26]

The net effect of the Army's self-questioning civic action programmes and collaborative victory against its larger neighbour – together with a revised constitution in 1998 that provided increased recognition to indigenous culture and identity – was an unofficial policy of military multiculturalism. Under this approach the Army tolerates indigenous languages, customs and beliefs within the garrison – so long as they do not impede military effectiveness. Officers also vigorously promote a national origin myth that emphasizes the roles of soldiers and Indians in the creation of the Republic, helping to obscure past discrimination.[27] Yet in practice this doctrine still seeks to eliminate indigenous men's 'backward ways' and make them 'productive members of society'. It therefore continues to perpetuate a legacy of paternal and asymmetrical relations, relegating *indígenas* to the role of junior (but essential) partners in national security. These shortcomings are overcome in large part by the fact that under this policy, *indígenas* can claim elevated status as citizens, soldiers and men through conscription, without entirely having to abandon their indigenous identities.

However, the Ecuadorean military's conditional acceptance of *indígenas* severely limits the value of this doctrine. The model is essentially a hybrid, blending elements of accommodation – whereby Indians are recognized as full citizens, and certain of their differences are allowed in the armed forces – and multiculturalism, which promotes deep and mutual understanding and respect.[28] While it is not bona fide multiculturalism, nor is it 'all smoke and mirrors', as it is quite a coherent set of ideas and practices. Yet neither is it a fully articulated belief system based on well-developed philosophical principles. Nevertheless, Ecuadorean officers seek to resolve the same basic tension between individual and collective rights as do liberal political theorists. In brief, this entails balancing in an equitable and practical manner, the individual and group rights as they apply to ethnic/minority groups and society as a whole.

Lacking solid conceptual bases, the Ecuadorean military's version of multiculturalism ignores or misinterprets academics' recommendations on how best to achieve this compromise. For example, Will Kymlicka recommends accepting 'some external protections for ethnic groups and national minorities', thereby limiting the majority's influence over them, while remaining 'very sceptical of internal restrictions' that limit members' freedoms and promote cultural purity.[29] The Ecuadorean Army inverts this equation, providing few external protections for *indígenas* vis-à-vis other groups (including the Army), and severely restricting, redefining and regulating their cultural rights within the armed forces. Under this sort of multiculturalism, conscription tolerates Indian men and some of their external differences, but seeks to replace their indigenous beliefs in return for the symbolic benefits of citizenship, manhood and military service. The result is that conscription makes indigenous recruits into *the Army's* image of multicultural citizens, while ensuring they remain monocultural (i.e. *mestizo*) men. Ironically, neither party seems conscious of these consequences, confirming the power of hegemony to naturalize what to an outsider appears strange.

While imperfect, this approach produced significantly better results than the Bolivian Army's policy until 2003. Indigenous–military tensions in Bolivia over the past two decades can be largely attributed to several factors. First, neo-liberal economic reforms – austerity, privatization, structural adjustment, commoditization of natural resources, etc. – imposed by the international community eventually curbed the hyper-inflation of the 1980s. Yet together with a downward turn in world tin prices, these measures led to high unemployment and public dissatisfaction. The Army was often called in to restore order, putting soldiers – including many indigenous recruits – in direct confrontation with a disgruntled citizenry. Second, many of the newly unemployed highland Indian miners migrated to the tropical lowlands, where they found work growing coca, a mildly alkaline plant with traditional (and legal) cultural, spiritual and medicinal uses in Bolivia. Coca, however, is also the raw material used in the production of cocaine, and Bolivia – together with Peru and Colombia – quickly became a key supplier of both.

As the United States militarized its 'War on Drugs' in the late 1980s and early 1990s, the Bolivian Army became an indispensable ally in combating the new scourge. Foreign political pressure, funding, training and equipment placed the Bolivian Army in increasing conflict with coca growers (*cocaleros*). Again, Indians found themselves on both sides, as soldiers in a new 'low-intensity war' and small-time producers of a legal plant often used for illicit ends. In the face of such pressure, *cocaleros* became increasingly organized, vocal and active, as indigenous groups in general had since the 1970s. Their collective pressure helped ensure that in 1994 the Constitution was modified to officially recognize Bolivia as 'multi-ethnic and pluri-cultural'. Such changes raised public expectations, so when administration after administration failed to enact these principles, some *indígenas* and *cocaleros* joined the political process, while many others turned to protest.

The 'Water War' in Cochabamba (April 2000), bloody highway blockades in the Altiplano (January 2003) and the police–military mêlée at the Presidential Palace (February 2003), were appalling in their own right, but also foreshadowed things to come. In October 2003, massive protests against a plan to export the country's natural gas through a port of arch-enemy Chile led to violent repression by the Army. An estimated ninety indigenous protesters died in what came to be known as 'Black October', which eventually led to the resignation of President Sánchez de Lozada. While tragic, these events also created unprecedented opportunities for transformation.

The newly installed president, Carlos Mesa, clearly signalled his intent to root out discrimination, reform national institutions and embrace the country's diversity as its strength. General Cesar López, Mesa's choice as Commander of the Army, embraced the president's vision and initiated an ambitious programme to reform the institution into a democratic and explicitly multicultural force. Specifically, he sought to fully incorporate *indígenas* into the officer corps, create programmes that better align the Army with civil society and culture, and reform conscription to replace traditional notions of exclusionary citizenship with a more inclusive concept.

If successful, these efforts will position the Army, one of the country's most powerful and effective but historically oppressive institutions, as a symbolic and participatory leader in long-term national-level reforms. The general first engaged Bolivian academics to create linkages with civil society and advise on the internal reforms. By November 2003 the Bolivian academic and military personnel recognized the requirement for additional conceptual expertise and practical experience to envision and guide such profound transformations. They turned to Canada, a leader in these areas, for assistance. Scholars from the Canadian Defence Academy and Royal Military College took the lead in this project, with funding support from Foreign Affairs Canada.[30]

Academic and military personnel from the two countries worked collaboratively, first to design institutional policies that supported then-President Mesa's government-wide goal of promoting 'strength through diversity'.[31] Eventually, they developed a specialized indigenous entry programme with the staff of the Bolivian Military College. The project engages indigenous community leaders to recommend candidates suitable to both parties for a six- to twelve-month preparatory course. This, in turn, assists native youth in making the necessary cultural and educational transitions, as well as creating a more supportive environment for diversity in general.[32]

The problem was not that indigenous people could not serve in the Army, or even the officer corps. The programme's detractors are quick to point out that the rank and file are primarily aboriginal and that many officer-cadets look or speak like indigenous people, or have indigenous surnames. They are correct; the root issue faced by the Bolivian Army was that individuals generally did not believe they could serve – and succeed – in the army as indigenous officers or soldiers. As a result, they temporarily abandoned or concealed their aboriginal identities through what the Bolivian commander once called 'cultural camouflage'. (The same could be said of many other countries' militaries and aboriginal soldiers.)

The pilot project aimed to reverse this sad legacy and establish a 'critical mass' of unequivocally indigenous cadets who are proud of their heritage and visibly successful. Yet accommodation is still new to Bolivia, so academics and officers from the two countries continue to collaborate on this project under the dynamic leadership of the new Army Commander, General Freddy Bersatti. Work is now under way to develop a broad spectrum of accommodation policies, from developing a multicultural curriculum to recognizing indigenous dress, customs, language and religion. Eventually, the entry programme could be replicated across the Bolivian military educational system and in other national institutions.

Dedicated entry programmes are an important component; however, without internal accommodation mechanisms as well, they are inadequate to achieve the sort of reform sought by Bolivian Army leaders. Focusing only on these sorts of programmes reduces native participation in the armed forces to a numerical 'problem' and encourages leaders to work to 'solve' it by focusing on recruiting initiatives. This does not address the conditions under which new and existing members serve. The unintentional results can be the creation of glass ceilings that increase indigenous representation at the bottom of the hierarchy but limit career advancement into more senior ranks. Such practices may also neutralize increased recruiting efforts by raising turnover rates, owing to dissatisfaction.

Therefore, in addition to establishing accommodation measures such as the indigenous officer entry programme, the Bolivian Army also sought to promote acceptance by launching a candid dialogue between the military and leaders of the country's indigenous majority. The commander publicly inaugurated this process in a highly symbolic ceremony at the Military College, where he personally welcomed indigenous representatives from every major group – twenty-four were originally invited, but the event proved so popular that sixty-six attended – in front of the entire Cadet Corps on parade. This sent a powerful message to the next generation of military leaders about acceptance of diversity.

A series of seminars and workshops were also organized for indigenous, academic and military representatives in March 2004, September 2004 and March 2006. The first event provided an opportunity to debate the future of indigenous–military relations, the second produced a concrete plan through active consultations,[33] and the third assessed efforts to date. Despite shortcomings and setbacks, indigenous leaders, the media, even the president praised the participants as frank and open, and the process as fair. They felt that the trust these events generated was a key factor in the groups' ability to avoid a recurrence of the violent clashes of October 2003. After years of inaction and strife, the Bolivian Army has decisively become a regional, perhaps world, leader in promoting military diversity by firmly embracing and enacting the principles of accommodation and acceptance.

Prospects for the future

Today, the Ecuadorean and Bolivian Armies are confronting similar challenges in strikingly different fashions. The Ecuadorean Army has accommodated indigenous demands by reconceptualizing the nation and conscription without significantly challenging the institution's stratified nature or core beliefs about *indígenas*' appropriate roles in national defence. In doing so, the institution has retained – with few exceptions[34] – the right to decide who is an Indian and how they may serve the nation. Despite the limitations of this utilitarian approach, it enjoys broad popular support from *indígenas* and soldiers alike, and has been sufficient to avoid widespread conflict between the groups. Yet perhaps for these same reasons, it has not contributed to broader multicultural reforms of the Ecuadorean state.

Conversely, in Bolivia, collaborative work between scholars, soldiers and Indians has created a programme that balances efforts to accommodate indigenous peoples at all ranks while generating acceptance for them as *indígenas*. This has required vision, political will, solid intellectual foundations, and a willingness to engage in frank, at times uncomfortable, exchanges. The results are far more transformative than those from Ecuador, altering both the composition and the culture of the Army. This was largely possible thanks to Army leaders' ability to link diversity to the institution's future success in core business – not just in the minds of generals and politicians, but for the public, indigenous and non-indigenous alike.

In the process, the Army has created a template for improving the nature of Bolivian governance in general. This was not lost on newly elected president (and former *cocalero*) Evo Morales, who declared in his January 2006 inaugural speech:

> In the National Army today there is no General Condori, General Willca, General Mamani, General Aima. To change this colonial state there will be spaces, debates and dialogues. We are obliged, as Bolivians, to understand each other to change this form of discrimination against the indigenous peoples.

Not surprisingly, shortly thereafter the Bolivian Air Force, Navy, National Police and other institutions declared their intent to implement similar programmes – and requested Canadian assistance to do so. Even more importantly, the administration has indicated a desire to codify and disseminate the guiding principles behind these efforts as a state policy, thereby ensuring increased consistency and scope in future reforms.

Similarly, efforts to link indigenous and other types of military diversity present opportunities to achieve more holistic and broad-reaching integration. Yet too often, gender and ethnicity are seen as discrete challenges, with indigenous women facing dual discrimination that has served to limit their entry to the armed forces. Moreover, while women are currently enrolled in both the Ecuadorean and the Bolivian Military Colleges, these programmes were designed as minimalist compliance efforts, accommodating the presence of women without ample efforts to ensure they felt like full members of the profession or that others accepted them as such. Consequently, neither army has successfully captured the administrative, cultural or policy lessons of these programmes to aid in indigenous integration efforts. Thankfully, there are signals that the Bolivian Army will address this deficiency by actively recruiting indigenous women to the Military College in 2006. Likewise, the National Police are keen to ensure that Afro-Bolivians are also included in their pilot project. When supported by rigorous research and assessment processes, such developments can significantly improve the results for the institution and civil society alike.

Indigenous integration in the Army also exerts a significant role on civil–military relations. For example, in Bolivia, senior military officers publicly voiced their commitment to honour the results of the December 2005 election, explicitly noting that they would assent to having an *indígena* as president. This is in part, though clearly not completely, a result of increased exchange and reduced conflict generated through public dialogues. While improving the military's relationship with the society it defends is generally a positive step, the results can occasionally be counter-productive. For example, consensus on key issues such as corruption led indigenous and military actors in Ecuador to unite and overthrow the elected president in January 2000.[35] It therefore seems that inclusive, transparent and horizontal relations such as those fostered in Bolivia since October 2003 – rather than the constrained, secretive and vertical sorts of contact that precipitated Ecuador's military–indigenous uprising – are healthier for democracy.

In the final analysis, Ecuador's military multiculturalism is too neat, and has been achieved without the key elements of participation and negotiation. While the results are an improvement over racist and exclusionary practices, they also mask persisting inequities and replicate historical asymmetries between the military and indigenous peoples. Thus, while they generate consensus on issues of national values and identity, these efforts are not truly emancipatory. The Bolivian case, on the other hand, has yet to produce complete consensus, but has resulted in tangible results and symbolic changes. Whether this will translate into military and indigenous actors' generalized support for policies and practices remains to be seen, as do the implications of these efforts for elsewhere in the region.

Notes

Thanks to Nicholas Gonzalez, Alan Okros and Loreta Tellería for their generous assistance in developing various aspects of this chapter. Any and all shortcomings, however, are solely the responsibility of the author.

Fieldwork in Ecuador was funded by the Social Science Research Council (SSRC), the Fulbright Commission and Syracuse University, whereas applied research in Bolivia has been supported by Foreign Affairs Canada and the Department of National Defence. While I am thankful for institutional and funding support, the views contained in this work are my own, and do not necessarily reflect the policies or opinions of the Governments of Canada, Bolivia or Ecuador.

1 H. Dietz and J. Elkin, 'The Military as a Vehicle for Social Integration', in H. Dietz, J. Elkin and M. Roumani (eds), *Ethnicity, Integration, and the Military*, Boulder, CO: Westview Press, 1991.
2 B.R. Selmeski, 'Aboriginal Soldiers: A Conceptual Framework', Kingston: Queen's-McGill University Press (in progress).
3 B. Loveman, *For la Patria: Politics and the Armed Forces in Latin America*, Wilmington, DE: Scholarly Resources, 1999.
4 L. Cabrera and E. Medina, *misión militar Chilena en el Ecuador: proyecto de ley orgánica militar*, Quito: Tipografía de la Escuela de Artes, 1902.

5 G.H. Andrade and T.A. Tapia, *Documentos para la historia de la Escuela Militar: 1830–1930*, Quito: Centro de Estudios Históricos del Ejército Ecuatoriano, 1991, p. 387.
6 M. Gnerre and J. Bottasso, *Del indigenismo a las organizaciones indígenas*, Quito: Abya-Yala, 1986, pp. 7–27.
7 P. Jaramillo Alvarado, *El Indio ecuatoriano*, Quito: Editorial Quito, 1922; S. Larrea, 'El servicio militar obligatorio conceptuado bajo los siguientes puntos, para que este de acuerdo con los intereses del estado', *El Ejército Nacional*, 8, 41, 1928, 294–304; R. Fontenebro, 'El servicio militar obligatorio', *El Centinela: Revista Militar*, 43, 111, 1929, 3–5; A. Albán Borja, 'La militarización del indio', *Bayardo*, 3, 1933, 56–58.
8 J.R. Quintana Taborga, *Soldados y ciudadanos: un estudio crítico sobre el servicio militar obligatorio en Bolivia*, La Paz: Programa de Investigación Estratégica en Bolivia, 1998, p. 35.
9 M. Prieto, *A Liberalism of Fear: Imagining Indigenous Subjects in Postcolonial Ecuador, 1895–1950*, Gainesville: University of Florida, 2003, p. 203.
10 C. Aillón Tamayo, 'El servicio de conscripción rural', *Revista de las Fuerzas Armadas Ecuatorianas* 3, 16–17, 1944, 20–23; C. Aillón Tamayo, 'Plan de trabajo para el servicio de conscripción rural', *Revista de las Fuerzas Armadas Ecuatorianas*, 4, 18–19, 1945, 26–30.
11 H.S. Klein, *Bolivia: The Evolution of a Multi-ethnic Society*, New York: Oxford University Press, 1982, pp. 188–194.
12 Quintana, *Soldados y ciudadanos*, 1998, pp. 54–73.
13 J.R. Quintana Taborga, 'El ambiguo mundo del servicio militar obligatorio', *Cuarto Intermedio*, 1997, 76–100.
14 D. Barrios de Chungara and M. Viezzer, *Let Me Speak! Testimony of Domitila, a Woman of the Bolivian Mines*, New York: Monthly Review Press, 1978; J. Nash, *I Spent My Life in the Mines: The Story of Juan Rojas, Bolivian Tin Miner*, New York: Columbia University Press, 1992.
15 B.J. Salguero, 'Conscripción agraria militar ecuatoriana "CAME"', *Revista Militar de las Fuerzas Armadas Ecuatorianas*, 5, 2, 1966, 52–55.
16 M.C. Barre, *Ideologías indigenistas y movimientos indios*, Mexico City: Siglo Veintiuno Editores, 1985.
17 Instituto Indigenista Interamericano, 'El Comercio, Quito: conscripción militar para campesinos', *América Indígena: Noticiero Indigenista*, 35, 1, 1975, 199–200.
18 N.E. Whitten, Jr, *Sicuanga Runa: The Other Side of Development in Amazonian Ecuador*, Urbana, IL: University of Illinois Press, 1985, pp. 41–44.
19 R. Stutzman, 'El Mestizaje: An All-Inclusive Ideology of Exclusion', in N.E. Whitten. Jr (ed.), *Cultural Transformations and Ethnicity in Modern Ecuador*, Urbana: University of Illinois Press, 1981, pp. 45–94.
20 R. Levoyer, 'Lo militares y el levantamiento indígena', in *Indios: una reflexión sobre el levantamiento indígena de 1990*, Quito: Instituto Latinoamericano de Investigaciones Sociales, El Duende, Abya-Yala, 1991, pp. 221–262.
21 I. Almeida, 'El movimiento indígena en la ideología de los sectores dominantes hispanoecuatorianos', in *Indios: una reflexión sobre el levantamiento indígena de 1990*, Quito: Instituto Latinoamericano de Investigaciones Sociales, El Duende, Abya-Yala, 1991, pp. 293–318.
22 B.H.E. Balladares, 'El movimiento indigenista en el Ecuador', *Revista de las Fuerzas Armadas del Ecuador*, 1991, 77.
23 J. Gallardo Román, 'Nación ecuatoriana o plurinacionalidad', *Revista de las Fuerzas Armadas del Ecuador*, 123, 1998, 19–25.
24 Comando Conjunto de las Fuerzas Armadas, Directive 90-06 for the development of cooperation and approximation programmes between the armed forces and the country's indigenous communities, Quito: Comando Conjunto, 1990.

25 Comando de la Fuerza Terrestre, Directive 04-92 for the functioning of the Civic Action and Forestation Companies (CACYF), Quito: Comando de la Fuerza Terrestre, 1992.
26 L.F. Duchicela, 'El conflicto como ejemplo de integración pluricultural en defensa de la integración nacional', *Revista de las Fuerzas Armadas del Ecuador*, 114, 1995, 74–76; L.B. Hernández Peñaherrera, *La Guerra del Cenepa: diario de un comandante*, Quito: Corporación Editora Nacional, 1998.
27 B.R. Selmeski, 'Who Is an Indian? The Politics of Identity in the Ecuadorean Military's 1994 "Indigenous Sanctions Programme"', paper given at the Canadian Anthropological Society (CASCA) Meeting, London, Ontario, 2004.
28 Selmeski, 'Aboriginal Soldiers'.
29 W. Kymlicka, *Multicultural Citizenship*, Oxford: Oxford University Press, 1995, p. 7.
30 Canada, 'Canadian–Bolivian Project on Military Diversity and Professionalization', 2005, www.cda-acd.forces.gc.ca/bolivia/engraph/home_e.asp.
31 N. Gonzalez, 'CF Supports Bolivia Army Reform', *The Maple Leaf*, 7, 34, 2004, 7.
32 L. Stone, *Promoting Diversity in the Bolivian Army: Setting the Stage*, Kingston, Ontario: Canadian Forces Leadership Institute, 2004; 'Firman un convenio para que 20 indígenas estudien en el Colmil', *El Diario*, 20 April 2005, www.cda-acd.forces.gc.ca/bolivia/engraph/media/doc/20-4.pdf. La Paz.
33 Combat Camera, CFLI/CDA Canadian–Bolivian Initiative (DVD) www.combatcamera.forces.gc.ca, Ottawa: Department of National Defence, 2004; N. Gonzalez, *Promoting Institutional Change: Perspectives from the Bolivia Initiative*, Kingston, Ontario: Canadian Forces Leadership Institute, 2004.
34 Selmeski, 'Who Is an Indian?'; B.R. Selmeski, '"We Are Family": National Metaphors and Popular Opinions of Conscription in Ecuador', in D. Last (ed.), *Challenge and Change for the Military: Social and Cultural Change*, Kingston, Ontario: McGill-Queen's University Press, 2004, pp. 101–115.
35 B.R. Selmeski, *Remarkable Images: Ecuador's Military–Indigenous Uprising*, New York: Latin American Video Archive (LAVA), 2000.

5 Diversity in the Brazilian armed forces

Celso Castro

Introduction

Examining the issue of diversity in the present-day Brazilian armed forces immediately gives rise to a serious problem: the sources. Not much public information is available on the matter, particularly regarding the Navy and the Air Force. Undoubtedly, the long authoritarian presence of the military, in power for twenty-one years – between 1964 and 1985 – has contributed to the lack of information on the internal demographics of the institution, something that could be viewed as a matter of 'national security'. It is important, however, also to ascribe part of the responsibility to civilian researchers, who have relegated the theme to a subordinate position. Because of the Marxist perspective dominant in the Brazilian social sciences during roughly the same period as the military regime, most social scientists viewed the armed forces as a kind of 'armed fist' of the bourgeoisie, within the wider context of class struggle. An interest in studying 'internal', 'structural' or 'organizational' aspects of the armed forces only gained ground, slowly, at the end of the military regime.[1] Besides, an emotional component certainly played a part in the hesitation to study an institution that so negatively affected the academic world, the social sciences in particular. This situation has undergone significant changes only in the past two decades, when, since the end of the military regime in 1985, the armed forces have lost much of their power in Brazil.[2]

Another difficulty in including a Brazilian chapter in a book about cultural diversity is the fact that the central themes being analyzed – gender and ethnicity – are especially difficult to approach in Brazil. The former is difficult because of the still recent incorporation of women into the armed forces; the latter because of specific characteristics of Brazilian society. I hope that the use of additional indicators – geographical background, religion, class and family background – will be helpful in offering a not-so-incomplete portrayal of the diversity of the Brazilian armed forces.

Before plunging into the central topics of this chapter, I will elaborate a bit more on some of the general characteristics of civil–military relations in Brazil. A brief historical overview provides a context for understanding even better the issues at hand.

The Brazilian military in politics and society: a very brief history[3]

From the end of the Paraguayan War in 1870 through the 1930s – a period that encompasses the end of Empire and the establishment of the Republic (1889) – the Brazilian military was an active social agent in the modernization of the country. Young Army officers educated at the Military Academy in Rio de Janeiro supported the affirmation of values and the implementation of practices linked to individualism, discipline, rationality, bureaucratic organization and merit. At that time, Brazilian society was still marked by a strong colonial heritage – slavery oriented and patriarchal – wherein aristocratic values and personal relations predominated. As an aftermath, the political progressiveness of the 1889 'scientific' youth and of the 1920s 'lieutenants' therefore inscribed on Brazil's political history the messianic vision that an 'enlightened' military group was able to 'save' the nation, in its name.

In the decades following World War II there were several military interventions in politics, leading to a twenty-one-year period of direct rule by the military, already mentioned, between 1964 and 1985. The military, and even some analysts, regarded such interventions as the manifestation of a type of 'moderating power' exercised by the armed forces. The idea was that the armed forces, representing the nation and going above and beyond the roles of the three traditional branches of government, had a right to intervene on the political scene when they judged it necessary, in order to solve institutional crises and serious political deadlocks. However, there was an important pattern to these interventions: political power was invariably and rapidly returned to civilians.

In this respect, 1964 broke away from this pattern: a genuine *military regime* was established, during which the military as an institution remained in power. But they did not rule alone. Powerful civilian groups – political, business, religious and middle-class sectors – stimulated, supported and collaborated with the successive military governments. Nevertheless, we can consider the regime to be military, because during the entire period the higher echelons of the armed forces were clearly the actors who kept political participation under strict control. When they were confronted by or became unsatisfied with even the limited political interplay that they allowed, the military governments acted several times in a highly authoritarian manner: shutting down Congress, amending the Constitution, refusing to respect the judiciary, deposing elected congressmen, mayors and governors from their offices, forcibly retiring public servants, revoking political rights, applying censorship, and exerting the most extreme forms of political repression (such as exile, prison, torture and even the outright assassination of members of the opposition). The legacy of this 'military regime', particularly with respect to the participation of military personnel in political repression, is still a sensitive issue in contemporary Brazilian historical memory.

The military regime in Brazil ended in 1985 after five successive military governments. The transition to civilian rule occurred under norms that still pertained to the authoritarian period. From the military's point of view, the Brazilian dictatorship ended after a successful transition to democratic rule. The transition was 'slow, gradualist and orderly', in the words of General-President Ernesto Geisel. However, in spite of the fact that the military coup of 1964 took place with high levels of support throughout Brazilian society, since the end of the military regime virtually no one public institution or important personality openly identifies with the dictatorship – not even those people who once supported the regime or benefited from it.

In the years of civilian democratic rule following 1985 (the so-called New Republic), members of the military were not punished for acts committed during the regime – an experience quite different from that in neighboring countries in the Southern Cone. While there was a retreat from the streets to the barracks, and the military accepted civilian democratic rule, they became socially more isolated, lost prestige and have yet to achieve a clearly defined social role. Nonetheless, the armed forces have tried to preserve for themselves the symbolic role of founders of the nation.

Gender

After this brief historical sketch of civil–military relations in Brazil, we will now turn to the social demographics of the present-day armed forces.[4] While in the course of this chapter most of the data presented refer to the officer corps and to the army, typically with regard to gender, the other two services are most conspicuous.

In Brazil the Navy was the pioneer in the integration of women into its institution, creating, in 1980, the Auxiliary Female Corps of the Navy Reserve (Corpo Auxiliar Feminino da Reserva da Marinha). This corps operated on the technical and administrative levels and allowed women to rank as far as *capitão de mar-e-guerra* (which corresponds to the rank of colonel in the Army).[5] In 1998 the Navy abolished the Female Corps and women could now, alongside men, join the intendancy, engineering and the health corps. The change gave women equal conditions in relation to men with regard to their access to courses and promotions, and they could ascend to posts as high as vice-admiral. A longstanding taboo had also been broken, since women obtained permission to participate in missions on board ship, and to join helicopter crews. Despite this progress, women are still not allowed to join the Navy Fleet, the most important career in the Navy. In March 2006, women constitute slightly over 4 percent of the Navy's military personnel, while they form some 26.5 percent of the civil workforce in the Navy.[6]

In 1981 the Air Force created the Female Corps of the Air Force Reserve (Corpo Feminino da Reserva da Aeronáutica), equally engaged in technical and administrative functions. Following the Navy's example, years later the Air Force began to grant women the right to progress up the hierarchy under the same conditions as military men. In 2006 the Brazilian Air Force comprised 74,318 military personnel, of whom 4,400 were women (6 percent).

It was in the Air Force that the greatest advances in terms of women's incorporation into the Brazilian armed forces occurred. In 1996 the first class with female cadets (seventeen of them) was initiated in the Air Force Academy within the Intendancy. This first class graduated in 1999. A few years later, women gained the right to become aviator officers. On 26 March 2004, female students of the Air Force Academy completed their first solo flight (i.e. without the instructor's presence) in the history of Brazil. Now women are able to ascend the hierarchy as far as the rank of brigadier.

The Army was the last service to allow female participation. In 1992 the first forty-nine women were enrolled in the Army Intendancy School, graduating as officers (first lieutenants) of the complementary staff of officers. Since they are not career officers, these women are not able to reach the highest level of the hierarchy. In 1997 they also started to be accepted in the Military Engineering Institute (IME) and the Health School of the Army. In December 2001 the number of women on active duty in the Army had risen to 2,170.[7] In February 2006, women numbered 3,896, or 2 percent of the Army's total strength of 177,899.[8]

The Army is still the service in which the participation of women is the most unequal in terms of career opportunity. The reasons pointed out as impeding wider integration are sometimes linked to the lack of funding for necessary adjustments (building female toilet facilities, for example), but it is clear that the main reason continues to be the prejudiced belief that the women cannot endure traditional activities in the military career for physical and psychological reasons.

In spite of the relative backwardness on the incorporation of women by the Army, the integration of women into the Brazilian armed forces generally has been successful. This is in line with the growing trend for women in Brazil to improve their social position during recent decades.

The situation is very different, however, regarding gays in the military, who encounter strong impediments of a moral nature. In D'Araujo's words:

> Women are generally viewed as individuals who need to be protected, both within and outside military quarters, logically implying their exclusion from certain activities considered to be too risky and rigorously disciplined and thus considered strictly masculine. Homosexuals, for that matter, are seen as victims of a behavioral deviation that threatens the technical and moral functioning of the military corporation or the military institutions as a whole.... The gay man is frequently associated with a threat to the tranquility of the troops, cadets and conscripts, for gay men are seen as not being able to control their impulses nor respect moral precedents that are conducive to a profession.[9]

An important step in changing this situation came in 2002, when the government put a bill before Congress on the subject of human rights, proposing the amendment of Article 235 of the Military Penal Code of 1969, which prescribes punishment for 'libidinous' practices, whether homosexual or not, in military facilities. The intention is to remove all mention of the words 'pederasty' and 'homosexual' from military rules and regulations. In any case, homosexuality is still a delicate subject in and out of the barracks.

Ethnicity

It is very hard to convincingly define ethnic boundaries within a deeply mixed society such as Brazil. The 182.1 million Brazilians in 2005[10] were the offspring of, broadly speaking, a threefold ancestry: first, a mixture of Europeans – especially the Portuguese, who colonized Brazil from the sixteenth century onwards, and immigrants from other European countries in the late eighteenth and early nineteenth centuries; second, Africans from different regions, who were taken there as slaves during the sixteenth to nineteenth centuries; and third, Amerindians from very different groups.[11] Rightly or wrongly, *mestiçagem* (the intermarriage of individuals of different 'races') is viewed positively by the population and is considered to constitute a distinctive feature of the national identity. Intermarriage is good in that it provides a real alternative to the rigid (and unscientific) classification of different human 'races'. It is bad that valuing interracial marriage can support and reinforce the false belief that there is no racial discrimination in Brazil. In spite of the ideological valorization of mixing, it is still true that the whiter one's skin, the more highly one is valued in the social and economic pyramid.

It is a rare and embarrassing experience for Brazilians to come across a field named 'race' while being asked to fill out official forms and paperwork during their everyday life as citizens. A field labeled 'ethnicity' would sound even stranger. With the exception of traditional indigenous populations – 0.4 percent of Brazil's total population[12] – a large number of foreign immigrants who migrated to Brazil at the end of the nineteenth century and the beginning of the twentieth, mostly Portuguese, Spanish, Italian, German, Syrian-Lebanese and Japanese, have been assimilated into Brazilian society over the course of a few generations. Unlike what happens in the United States, for instance, it is rare to find a Brazilian citizen defining him- or herself as Italian-Brazilian or Japanese-Brazilian.

Brazil has no official statistics regarding the presence of different 'races' or 'ethnicities' in the armed forces. Members of the military share the national ideology that views Brazilian nationality as a result of the fusion of three 'races': white, black and indigenous. A nice illustration is offered by Army Day, celebrated on 19 April, the day of the first battle of Montes Guararapes in 1648, which was decisive in driving away the Dutch, who had invaded northeast Brazil. The narrative associated with this episode always emphasized that victory over a richer and more powerful enemy was obtained by the union of the 'three races', represented by the black, white and indigenous troops who fought in the battle.[13]

Throughout the whole republican period (which began in 1889) there has been no experience of formal racial segregation within the military units. During World War II, when a Brazilian expeditionary corps fought in Italy against the Germans, meeting the racially segregated American troops was a source of discomfort. It is fairly common in the Army to hear the expression 'here, everyone is green', which is a reference to the color of the uniform. In the Navy, traditionally there was an informal barrier against the membership of blacks and mulatos, due to the prohibitive cost of their complete set of uniforms (*enxoval*), which they were required to purchase on being accepted to the Naval School. Recent decades, however, have witnessed visible changes in this state of affairs.

McCann, using yearbook photos from the Preparatory School and AMAN – the Military Academy of the Agulhas Negras – and informal conversations with officers, estimated an enrollment of about 20 percent Negro/mulatto personnel for the Army officer corps as a whole from the mid-1970s onward, and perhaps as much as 40 percent of the students of the Preparatory School.[14] It is my personal impression, based on my research experience and familiarity with the military,[15] that the degree of racial prejudice seems to be *lower* within the armed forces, particularly in the Army, than in Brazilian society generally. In the military it is much easier, and customary, for a black man to have command over a white man than in other careers or institutions.

Geographical background

Considering the distribution of military personnel according to their geographical background can be important in a country with continental dimensions like Brazil. Table 5.1 relates national census data to the birthplace of cadets belonging to AMAN, the sole institution of higher learning from which the officer corps in the Brazilian Army is taken.

Comparing the birthplace of AMAN cadets with the average national population over forty years, we note that there is a noticeable convergence between

Table 5.1 Regional background of Brazilian officer-cadets

Region	1960		1980	2000
	Brazil	AMAN[a]		

70 C. Castro

the two. The slight over-representation of the Southeastern region can be attributed to the fact that it is in this region that both the Preparatory School for Cadets of the Army (EsPCEx) and AMAN are located. Looking more closely at this region, however, we note an over-representation in the number of cadets born in Rio de Janeiro State – where AMAN is located – compared to the Brazil's most populated state, Sã Paulo. Anyway, speaking in broad terms of geographic diversity, the Army is not exaggerating in its claim to be a national institution. This aspect is reinforced by the intense geographic mobility of the officers during their career, thought of by the institution as a means of bolstering national sentiment.

The same cannot be said of the other services. Between 1998 and 2002 the Navy Academy enrolled 90 percent of its cadets from the Southeast (mostly from Rio de Janeiro State, where the Navy Academy is located). In 1992, 58.7 percent of the cadets of the Air Force Academy were born in Rio de Janeiro State. So, we can say that the Navy and Air Force officer corps are much more *carioca* or *fluminense* (born, respectively, in the city or in the state of Rio de Janeiro) than nationally representative in their composition, compared to the Army.

Religion

Brazil is the largest Catholic country in the world in terms of the number of believers. This affiliation, which is preponderant throughout the whole of Brazilian history, has diminished in recent decades, owing to the increasing presence of evangelical sects. The proportion of Catholics and Evangelicals among AMAN cadets, expressed in Table 5.2, practically mirrors the national average, including the decrease in numbers of Catholics.

One must note the over-representation of Spiritism – practiced by Kardecist Spiritists – which is a long tradition among the military in Brazil, and also present in the Navy. Among the cadets of the Navy School, Spiritists make up some 12 percent, Evangelicals 17 percent and Catholics around 60 percent.

Table 5.2 Religious background of Brazilian officer-cadets

	1980		1991	2000
	Brazil	AMAN		

Let me offer a short explanation concerning Spiritists, who are generally from middle-class origins. Spiritism is a religious and philosophic Christian doctrine established in France in the mid nineteenth century by Allan Kardec. Its practitioners believe in communication with the spirits of the dead, and reincarnation. Nowadays it is a quite stable religious denomination, especially strong in Brazil and, to some extent, in the Philippines. Syncretism of Spiritism and Afro-Brazilian cults such as Umbanda is a common religious and cultural experience in Brazil. The Cruzada dos Militares Espíritas (Crusade of the Military Spiritists), created in 1944, is believed to have up to 5,000 participants.

Social background

In the case of the Brazilian Army we possess a historical series of data about the social background of the officer corps. Alfred Stepan, researching AMAN at the end of the 1960s, obtained data for the 1940s and 1960s.[16] Stepan himself recognizes the existence of many problems related to these data, notably the lack of detail in the information, which appears in broad categories. Professions such as, for instance, 'military' and 'civil servant' were included in the category of 'middle class', even though the former may embrace any post in the military hierarchy, from general to private soldier, and the latter may refer to either a highly ranked person in the bureaucracy or a porter in a public building. Nevertheless, according to Stepan, the data allow general tendencies to be observed.

Stepan grouped professions and occupations into a few 'classes': 'Traditional Upper' (farmers, bankers, magistrates, doctors, lawyers, etc.); 'Middle' (*military*, civil servants, traders, teachers, etc.); 'Skilled Lower' (industrial workers, railway workers, drivers, craftsmen, etc.); and 'Unskilled Lower'. Lastly, there is the 'unknown' category, which includes orphans, the non-classified and undeclared professions. Table 5.3 offers a simplified reproduction of the data presented by Stepan.

Table 5.3 Social background of Brazilian officer-cadets

Class	Years	
	1941–43	*1962–66*
Traditional upper	19.8	6.0
Middle	76.4	78.2
Skilled lower	1.5	8.6
Unskilled lower	2.3	0.4
Unknown or unclassified	–	6.7
	($N=1,031$)	($N=1,176$)

Source: A. Stepan, *The Military in Politics* (see note 16). As for the category 'unknown or unclassified', McCann has observed that 'AMAN officials believe that those cadets who do not list father's profession act from embarrassment, and they regard these cadets as being from lower-class families' (McCann, 'The Military'; see note 14).

By observing the data, Stepan and Barros[17] underpinned the *decrease in status* of the social background of cadets between the 1940s and 60s, with an increase in individuals belonging to the 'lower class'. However, the mass of the recruiting pool remained in the 'middle class'. By the way, Army officers always enjoyed having this representation as 'middle class', often invoking it to justify the self-ascribed role as 'mediators' in Brazilian politics and in the stances of armed intervention. In the past three decades, however, it has become common to listen to military officers speaking about a 'proletarianization' process going on in the recruitment of AMAN's cadets.

The superficiality and inconsistency of the data exposed above led Stepan to seek another indicator for the social background of the cadets: the schooling of the parents of cadet's who enrolled in AMAN from 1963 to 1965. The last grade concluded was: higher education 29.6 percent; high school 9.5 percent; elementary and middle school 60.9 percent. These results led the author to state that enrollment in the Military Academy was a clear means of increasing social mobility, an indication that the center of gravity resided in the lower middle class.

As mentioned previously, this information is still very imprecise, primarily because of the lack of definition of what these 'classes' really mean. However, there is another data arrangement that is more precise and relevant. It refers to the percentages of the cadets whose parents are civilians or soldiers, respectively. Table 5.4 shows an elaboration of Stepan's table, to which I have added data for the years 1984–85 and 2000–02, all obtained from AMAN. Thus, the periodicity element of approximately two decades between the data is maintained.

Two important observations can be suggested in relation to Table 5.4. First, there is a high percentage weight of the military component in terms of the 'class' distinction formulated by Stepan. The question asked about the military's socioeconomic background, therefore, was corrupted right from its very inception. The high percentage of sons of the military raises the issue of how meaningful social background (in terms of 'social classes') is. It looks as though that the important thing was not to detect the 'socioeconomic class' background, but rather to ask why there are so many sons of military parents. The concern about determining officers' social class origins would be more appropriate to a theoretical perspective that focused on the position they occupy in the social struc-

Table 5.4 Cadets' parents' background

Parents	1941–43	1962–66	1984–85	2000–02
Civilian	78.8	65.1	48.1	54.6
Military	21.2	34.9	51.9	45.4
	(N=1,031)	(N=1,176)	(N=812)	(N=1,274)

Source: AMAN. The category 'unknown or unclassified' has been left out. For the last period, parents with a military background include five mothers.

ture – thus, present position would be associated with the 'class', 'segment' or 'stratum' from which the officers originally came. In another perspective, data about social background would not be as relevant as the significance of military professional socialization, and of the social networks officers have, for the understanding of their behavior and worldview. Here the focus of analysis would then be the 'internal', 'organizational' or 'structural' aspects of the military institution. Since, as a result of a persisting trend, nearly half of the Army cadets in the 1980s and 1990s – from whom will come the generals of 2020 – have military fathers, the study of the professional socialization of the military and the sociability during the career would be privileged as analytical foci.[18]

If we keep in mind that, as has been said before, the 'military' category embraced all hierarchical gradations, its placement in the 'middle class' may generate significant distortions. Alternative data that allow for a detailed outlook are unfortunately not available for the time period covered by Stepan. Barros, however, has presented information relating to 1970,[19] to which I add other data referring to 1985 and to 2000–02.

What can be concluded from Table 5.5 is that, at least for Stepan's data regarding the time period between 1962 and 1966, the military should have been allocated not to the 'Middle' class, but rather to the 'Skilled Lower', since the percentage of Lower Military is larger than that of the superior officers. Thence the center of gravity of recruitment, which Stepan imagined to be displaced over time from the 'middle class' to the 'lower middle class' would be confirmed if we consider the Lower Military to be a part of the 'lower middle class', or it would result in fact in a clearer shift to the 'Skilled Lower' if we place the Lower Military in the 'Skilled Lower' – whatever these categories may mean.[20] Since the 1970s there seems to have been a slow convergence trend among the percentages of the two groups, although there is still a clear predominance of sons of Lower Military.

The situation in the other services is different.[21] In 1975, 67 percent of first-year Naval School students' parents were Upper Military. In 1985 and 1990 the percentages were, respectively, 75 percent and 59 percent. Between 1995 and 1998 it decreased to 50 percent, and in 1998–2002 to 45 percent. The data clearly show a tendency towards a decrease in the proportion of Upper Military

Table 5.5 Cadets' parents' background: upper versus lower military (%)

Parents	1970	1985	2000–02
Upper military	28.5	31.9	41.9
Lower military	72.5	68.1	58.1
	($N=144$)	($N=210$)	($N=575$)

Source: Barros, *The Brazilian Military* (see note 17) and author. I am following the classification of McCann: upper military stands for majors through to colonels and would be AMAN's graduates. Lower military are soldiers, corporals, sergeants, (sub-)lieutenants and captains. The latter three ranks would be former sergeants who attained their status outside AMAN and never served as commanders (McCann, note 14).

during the second half of the 1980s. The neighborhood of cadets resident in the city of Rio de Janeiro (86 percent of all Navy cadets) also shows the same tendency, since between 1998 and 2002 only 7 percent lived in the South Zone (where most Upper- and Middle-class *cariocas* live), compared to 31 percent in the North Zone (mostly home to the Lower Middle class) and 39 percent in the *Baixada* or suburban areas (mostly Lower-class residents).

As for the Air Force, the data I obtained in relation to the rank of cadets' parents in 1992 are the following: Upper Military 42.9 percent, Lower Military 57.1 percent. So, we can state that in the Navy and possibly the Air Force the percentage of Upper Military's sons is much larger than in the Army.

The second important observation that can be articulated refers to an increasing trend in endogenous recruiting, which occurred until the mid-1980s. The percentage of sons of military parents in the Navy and Air Force is much lower than for the Army. Among the cadets of the Escola Naval (EN) in 1975, 28 percent had military parents, a percentage that remained stable in the 1980s and has been decreasing since then: 29 percent in 1985, 27 percent in 1990, 24 percent in 1995–98 and 20 percent in 1998–2002. In the Air Force Academy, out of the 748 cadets enrolled in the four years of the course in 1992, only 161, or 21.5 percent, had military parents. Therefore, in the Navy and the Air Force exogenous recruitment (among sons of civilian parents) dominates.

Reflecting during the 1970s on the endogenous trend of officer recruitment of the Army, Stepan and Barros arrived at similar general conclusions. They highlighted the reinforcement of corporate ties and the loosening of ties with civilians in the period before and after the military coup in 1964, which deepened the chasm between civilians and the military. Endogeny and the specificity of the socialization pattern were processes that mutually reinforced each other. Even though no cause–effect relationship has been demonstrated, one could propose a connection between the closing of the political arena and more endogeny.

The data for 2000–02 suggest a slow reversal of this tendency, yet still with a high percentage of military parents – more than double that in the 1940s. One of the characteristics of complex societies is the coexistence of diverse social circles among which the individual moves. Through this trajectory, the individual becomes more and more a unique being, distinct from other individuals.[22] However, Army officers' social experience in recent decades has been characterized by the concentration of many spheres of life within the same social circle: they spend their career years in military clubs, military schools, military villages, etc. When we add to this the fact that a large percentage of Army officers are sons of military personnel, the risk exists that the boundary between the military world and the civilian world may become very sharp.

Conclusions

There are differences between the organizational cultures of the three Brazilian armed forces – in Charles Moskos's terms, the Army resembles the 'institutional' model (emphasizing divergences in relation to civil society), while the

Navy and Air Force are closer to the 'occupational' model (emphasizing convergences in relation to civil society). In general, however, we can observe a series of elements common to the three services that point to reasonable tolerance towards diversity, except regarding homosexuality.

In spite of data scarcity, particularly in reference to the Lower Military, we can observe that the Army officer corps is quite close to the national average in relation to religious affiliation and birthplace, and that the racial issue is at least no worse than the average within Brazilian society. We can also observe that, in the case of the Army – and I believe that this statement is valid for the other two services – a military career is an important means of social mobility in Brazilian society, and, moreover, guided by meritocratic rules.

As for the integration of women, the Navy and the Air Force have introduced changes in the past decade in order to provide women with conditions equal to those for male military officers, and it is expected that the Army will end up doing the same. Although most Brazilian women are still restricted to the exercise of administrative or technical functions, there is good reason to believe that, over the following years, there will be an increase in female participation in career-based activities in the military.

On the other hand, a potentially problematic point has been, in the past four decades, the great concentration of military offspring in the Army officer corps. That points to the reality of an institution with uneasy links to society[23] and to a still unaccomplished agenda of better civil–military relations. And, of course, to the need for more research on the military in Brazil.

Notes

1 For a summary of these theoretical perspectives, see Edmundo Campos Coelho, 'A instituiõ militar no Brasil: um estudo bibliogrfico' (The military institution in Brazil: a bibliographical study), Rio de Janeiro: *BIB* no. 19, 1985.
2 See Celso Castro and Maria Celina D'Araujo (eds), *Militares e política na Nova República* (Military and Politics in New Republic), Rio de Janeiro: FGV, 2001.
3 For much more detailed views on the topics adressed within this short history of civil–military relations, see a number of my own publications: C. Castro, 'The Army as a Modernizing Actor in Brazil, 1870–1930', in P. Silva (ed.), *The Soldier and the State in South America: Essays in Civil–Military Relations*, New York: Palgrave, 2001, pp. 53–69; C. Castro, 'The Military and Politics in Brazil, 1964–2000', in K. Konings and D. Kruijt (eds), *Political Armies: The Military and Nation Building in the Age of Democracy*, London: Zed Books, 2002, pp. 90–110; C. Castro, 'Commemorating the "Revolution": historical memories of the Brazilian military', paper presented at the International Symposium 'The Cultures of Dictatorship: Historical Reflections on the Brazilian Golpe of 1964', the University of Maryland and Archives II, College Park, MD, 14–16 October 2004.
4 Unless explicitly stated, all the data used in this chapter were obtained at the military academies. I would like to thank Nathalia Levy for helping me with the organization of the data.
5 See the Armed Forces home pages (www.exercito.gov.br, www.mar.mil.br and www.aer.mil.br) and the chapter by M.C. D'Araujo, 'Mulheres, homossexuais e Forças Armadas no Brasil' (Women, homosexuals and armed forces in Brazil), in C.

Castro, H. Kraay and V. Izecksohn (orgs.), *Nova história militar Brasileira* (New Brazilian military history), Rio de Janeiro: Ed. FGV e Bom Texto, 2002.
6 Source: Brazilian Navy. I would like to thank Antônio Jorge Ramalho for providing this information.
7 M.A. Damasceno Vieira, *Presença feminina nas Forças Armadas* (Female presence in the armed forces), Braslia: Cânara dos Deputados, 2001, www2.camara.gov.br/publicacoes/estnottec/tema3/pdf/112264.pdf.
8 I would like to thank Antônio Jorge Ramalho for this information.
9 D'Araujo, 'Mulheres, homossexuais e Forças Armadas', pp. 444–450.
10 Demographic Census of IBGE (Brazilian Institute of Demographics and Statistics), 2006.
11 The genetic study of S. Penna, 'Retrato molecular do Brazil, versã 2001' (Molecular Portrait of Brazil, version 201), in *Homo Brasilis*, Ribeirã Preto: Funpec, 2002, examined maternal mitochondrial DNA from Brazilians considered 'white' and identified their composition as having 33 percent of genes that were of 'Amerindian' ancestry, 28 percent 'African' and 39 percent 'European'.
12 Data from the demographic Census of IBGE in 2000. It is very difficult to estimate the size of the Amerindian population at the beginning of the sixteenth century, when the first Europeans came to Brazil. Figures generally vary from one to five million. The population was decimated by epidemics and slavery in the following centuries, but the Amerindian genetic and cultural heritage is still very much present in Brazilian society.
13 C. Castro, *A invenção do Exército Brasileiro* (The invention of the Brazilian Army), Rio de Janeiro: Jorge Zahar, 2002.
14 F.D. McCann, 'The Military', in M.L. Conniff and F.D. McCann (eds), *Modern Brazil: Elites and Masses in Historical Perspective*, Lincoln NE: University of Nebraska Press, 1989, p. 69.
15 Besides the biographical fact that my father was an Army officer and, because of that, I lived in several military facilities during my youth, since 1987 I have conducted approximately 300 hours of recorded interviews with about 100 military personnel, including cadets (students at the Brazilian Army Officer Academy), officers on active duty during the authoritarian military regime (1964–85) and military leaders of the 'New Republic'.
16 A. Stepan, *The Military in Politics: Changing Patterns in Brazil*, Princeton, NJ: Princeton University Press, 1971.
17 A. de Sousa Costa Barros, 'The Brazilian Military: Professional Socialization, Political Performance and State Building', PhD dissertation, University of Chicago, 1978, pp. 60–62.
18 C. Castro, *O espírito militar* (The military spirit), 2nd edn, Rio de Janeiro: Jorge Zahar, 2002.
19 Barros, 'The Brazilian Military', p. 64.
20 McCann places the Lower Military into the Lower Middle class. In doing so, he found for the period 1982–85 totals of 4.5 percent for the Upper Class, 34.4 percent for the Middle, 45.9 percent for the Lower Middle and 15.2 percent for the Skilled Lower. McCann, 'The Military', p. 68.
21 Data about Navy and Air Force come from C. Castro, 'A origem social dos militares: novos dados para uma antiga discussã' (The social origins of the military: new data for an old discussion), *Novos Estudos CEBRAP* no. 37, 1993, 225–231. I managed to update the information.
22 For this sociological perspective, see Georg Simmel, *On Individuality and Social Forms*, Chicago: University of Chicago Press, 1971, particularly ch. 18, 'Group Expansion and the Development of Individuality.'
23 McCann, 'The Military', p. 75.

6 Diversity in the South African armed forces

Lindy Heinecken

Introduction

Five years ago, when I wrote on managing diversity in the South African armed forces, it was with the firm belief that I was part of a rainbow nation that had survived a peaceful democratic transition and that, at last, South Africans would be a united nation.[1] However, as the 'Madiba magic' of President Mandela started to wear off and President Thabo Mbeki came to implement the newly forged affirmative action and black empowerment policies of government, so new tensions and insecurities emerged.[2] Although most Whites expected and accepted change, few anticipated that it would be so rapid and radical – or that one's future would be influenced, once more, by the colour of one's skin. Nowhere was this more apparent than in the armed forces, where the transformation of the new South African National Defence Force (SANDF) marked the transfer of power from white to black officers in the years to come.

Under the African National Congress (ANC)-led government, an assertive affirmative action strategy was implemented to correct the past social and economic inequalities caused by years of colonialism and apartheid. Hence, it is appropriate to commence with an overview of the demographic profile and disparities in terms of education, unemployment and poverty that justify government's black empowerment initiatives. The various legislative provisions instituted to correct existing racial and gender imbalances are highlighted, and it is shown how they relate to the management of diversity within the armed forces. However, the SANDF faced the challenge not only of restoring past racial imbalances in the SANDF, but also of integrating seven different armed forces.

More so than cultural differences and gender equality, race and former force continue to be a source of tension, giving rise to problems of morale, discipline and cohesion in the SANDF. To deal with this, various management of diversity programmes have been instituted through the years. While these initiatives are commendable and have sensitized members to the difference among them, it is concluded that formidable challenges still lie ahead in managing diversity in the SANDF.

Brief overview of South African society

South Africa has a multiracial and multi-ethnic population of 46.5 million of which black Africans (hereafter Africans) constitute 79.3 per cent, Whites 9.5 per cent, Coloureds 8.8 per cent and Indians 2.4 per cent. Africans belong to nine ethnic groups – Zulu, Xhosa, Pedi, Sotho, Tswana, Tsonga, Swazi, Ndebele and Venda – each with its own language.[3] Currently there are eleven official languages, of which the most important in terms of number of speakers are Zulu (23.8 per cent), Xhosa (17.6 per cent) and Afrikaans (13.3 per cent). Although English is the official language of communication, only 8.2 per cent of the population cite this as their first language.[4]

Vast disparities still exist in terms of the educational profile of the different racial groups. Under apartheid,[5] schools were segregated, and the quantity and quality of education varied significantly across racial groups. Although a long and arduous process of restructuring the country's educational system began in 1994 (and is ongoing), only 5.2 per cent of black Africans have any form of higher education, compared to 29.8 per cent of Whites, 14.9 per cent of Indians and 4.9 per cent of Coloureds. At present, 22.3 per cent of Africans have no formal schooling whatsoever, compared to only 1.4 per cent of Whites.[6] As a result, South Africa is saddled with millions of illiterate and poorly trained citizens.

Consequently, unemployment is highest among those from the previously disadvantaged groups (Africans, Coloureds and Indians – collectively referred to as Blacks), and is worsening. Unemployment has increased by more than three million since 1994 and is estimated to be around 30–40 per cent. It is highest among Africans at 48.8 per cent, followed by Coloureds at 29.4 per cent, Indians at 20.7 per cent and Whites at 7.6 per cent.[7] An estimated 62 per cent of Africans live below the poverty line and, along with Brazil, South Africa is regarded as the most unequal society in the world, with a Gini coefficient of around 0.6. Clearly, such inequality not only is destabilizing, but, when coupled with race and the impact of HIV/Aids, makes for a particularly noxious force.[8]

Currently South Africa is the country most affected by HIV, with five million HIV-infected individuals. Twenty per cent of the 15- to 49-year-old population is infected and in parts of the country more than 35 per cent of women of childbearing age are infected. Approximately 40 per cent of deaths are Aids-related. Without effective prevention and treatment, 5–7 million cumulative Aids deaths are anticipated by 2010 (with 1.5 million deaths in 2010 alone). Estimates are that the epidemic could cost South Africa as much as 17 per cent in GDP growth by 2010. All these social issues have a marked impact on the SANDF and, directly or indirectly, on the management of diversity.

Legislative framework

Given the above, various laws have been passed to correct past racial imbalances and to protect citizens from unfair discrimination, including laws relating to HIV/Aids. The SANDF is obliged to implement corresponding policies and to abide by these political imperatives.

Constitutional and legal provisions

In South Africa the principle of equality is firmly entrenched in the Bill of Rights in the Constitution of the RSA, 1996 (hereafter the Constitution), Section 9(3), which states that unfair discrimination, whether direct or indirect, on any grounds of 'race, gender, sex, pregnancy, marital status, ethnic or social origin, colour, sexual orientation, age, disability, religion, conscience, belief, culture, language and birth is prohibited'. Section 9(2) states: '[T]o promote the achievement of equality, legislative and other measures designed to protect or advance persons, or categories of persons, disadvantaged by unfair discrimination may be taken.'[9]

To address the racial imbalances in the public service, Section 195(i) of the Constitution specifically states that the 'public administration must be broadly representative of the South African people, with employment and personnel management practices based on ability, objectivity, fairness, and the need to redress the imbalances of the past to achieve broad representation'. The implication of these constitutional provisions is that racially biased legislation may be promulgated to advance previously disadvantaged persons (including Blacks, women and the disabled) and that state departments are obliged to take active steps to advance those so classified.

Specific laws relating to racial and gender equality have been promulgated in recent years. These include the Employment Equity Act, No. 55, 1998, which places punitive sanctions on employers who fail to achieve equitable racial representation of Blacks in the workplace. This Act excludes the SANDF because of the unique functions it performs, but includes civilians employed in the Department of Defence (DOD). However, the Promotion of Equality and Prevention of Unfair Discrimination Act 2000, which came into operation on 16 June 2003 and which prohibits unfair discrimination on seventeen grounds, including race, gender and disability, does apply to the SANDF.

To ensure representivity[10] at all levels within the public service, an array of White Papers have been promulgated since 1994, such as:

- The White Paper on the Transformation of the Public Service, 1995, sets out a comprehensive framework for change in line with the above-mentioned constitutional provisions, and specifically states that affirmative action (AA) on behalf of previously disadvantaged groups is the main foundation of a non-racist, non-sexist society.
- The White Paper on Human Resource Management in the Public Service, 1995, provides a policy framework for the development of human resource management practices geared towards economic and social transformation.[11]
- The White Paper on Affirmative Action in the Public Service, 1998,

outlines the steps national departments and provincial administrations must take to develop and implement their AA programmes.

Department of Defence policies

The constitutional provisions, together with the above-mentioned White Papers affecting the transformation of the public service, have a profound impact on racial and gender representivity in the DOD. Taking cognizance of these provisions, the White Paper on Defence, 1996, affirmed that 'to secure the legitimacy of the armed forces, the Department of Defence (DOD) is committed to the goal of overcoming the legacy of racial and gender discrimination', and will seek to create a defence force that is professional, efficient, effective and broadly representative.[12]

In line with the White Paper on the Transformation of the Public Service, Parliament requested that the Minister of Defence oversee the design and implementation of an affirmative action (AA) and equal opportunity (EO) programme to identify and eliminate discriminatory practices and attitudes in the Defence Force.[13] In June 1996, after wide consultation, the Minister of Defence approved the establishment of an EO and AA Chief Directorate, tasked among other things with monitoring and evaluating the progress of EO and AA programmes in the DOD. Major General P.R.F. (Jackie) Sedibe, the first black women general in the SANDF, was appointed as the Chief Director EO, and the Chief Directorate EO and AA formally commenced its activities at the beginning of 1997.[14]

In June 1998 the Chief Directorate released the DoD's first EO and AA policy. This policy was updated in 2002 and outlines the DOD's approach towards racial discrimination, employment equity for women, gender and sexual harassment, sexual orientation, religious and language accommodation, and the management of diversity.[15] A policy was also promulgated in 2002 on the 'Prevention and Eradication of All Forms of Gender Based Violence'.[16] Besides these policies, the DoD's Civic Education Programme, implemented in 1995, aims to instil respect among DOD members for the core values of a democracy, and until recently included a specific chapter on the management of diversity.[17]

Challenges associated with the management of diversity

Up until 1993, the South African Defence Force (SADF) was a conscript force consisting mostly of White males, with a large White part-time force component. Disenfranchised Blacks were not conscripted, but could volunteer to serve in the military. Whereas initially these volunteers served only in non-combat positions, they were increasingly drawn into the SADF for service in operational areas, but remained junior partners in defence, even though proportionately they were deployed in front-line roles to a much greater extent.[18] Women were not conscripted, but served in supportive roles from 1970 onwards, their functional role being limited to support branches, such as finance, personnel, logistics, intelligence, medical services and welfare.[19]

With the formation of the SANDF in 1994, the racial and gender composition has changed dramatically. Not only did the demographics change, but as the years progressed, so too did the power relations between the various forces integrated to form the new SANDF.

Integrating former enemies

One of the most important processes associated with the peaceful transition to democracy in South Africa was the need to integrate the seven different military forces. The forces integrated to form the new SANDF on 27 April 1994 included the former statutory forces consisting of the SADF and the Transkei Defence Force (TDF), Bophutatswana Defence Force (BFD), Venda Defence Force (VDM) and Ciskei Defence Force (CDF), collectively known as the TBVC forces, and the former non-statutory revolutionary forces of Umkhonto we Sizwe (MK), the liberation army of the ANC, and the Azanian People's Liberation Army (APLA) of the black consciousness Pan African Congress (PAC). Included later was the KwaZulu Self Protection Force (KZSPF) of the Inkhata Freedom Party.

By April 1998, on completion of the integration process, the SANDF consisted of 16.0 per cent, former Umkhonto we Sizwe, 6.7 per cent Azanian People's Liberation Army, 8.8 per cent TBVC, 53.2 per cent South African Defence Force, 2.4 per cent KwaZulu Self Protection Force and 13.0 per cent SANDF 'propers' (recruited after 1994).[20] As can be seen from Table 6.1, the current breakdown of the SANDF by former force as on 15 November 2005 out of a total force strength of 60,849 (excluding civilians) consisted of Umkhonto we Sizwe 14.3 per cent, Azanian People's Liberation Army 6.6 per cent, TBVC 7.4 per cent, South African Defence Force 38.4 per cent and SANDF 'propers' 33.3 per cent.[21] Although the integration process is now complete, this does not imply that the SANDF is a unified force. The past years have seen numerous

Table 6.1 Composition of the South African National Defence Force by former force and comparative representation in rank group brig.-gen. and above at 15 November 2005 (%)

	Composition	Rank profile
MK	14.3	30.8
APLA	6.6	9.2
TBVC	7.4	7.4
SADF	38.4	52.5
SANDF	33.3	0.0

Source: Department of Defence personnel statistics.

Note
MK = Umkhonto we Sizwe; APLA = Azanian People's Liberation Army; TBVC = the former Transkei, Bophutatswana, Venda and Ciskei defence forces; SADF = South African Defence Force; SANDF = South African National Defence Force.

reports on racism and tension within the ranks, prompting a ministerial investigation into the integration process.[22]

In September 1999 a disgruntled former APLA member shot dead seven White fellow soldiers as well as a civilian clerk at a military base in Tempe in Bloemfontein.[23] In July 2000 an African lieutenant killed his White company commander at 7 South African Infantry Battalion and a month later an African naval rating shot his White watch officer aboard a naval minesweeper, SAS *East London*.[24] There was one more racial killing during 2001, when a black corporal serving on the peacekeeping mission in Burundi shot and seriously injured a white officer. The Defence Minister stated publicly that racism was a likely motivating factor in the past killings.[25]

Although such incidences appear to be on the decline, many still feel that the old SADF culture still prevails and that the non-statutory forces have merely been absorbed into the SANDF. In some respects this is true. Like the former SADF, the SANDF (like any other conventional force) is a corporate organization, authoritarian, rigidly hierarchical, with set rules and regulations, allowing little room for initiative and freedom of action. In contrast, the revolutionary forces operated in small independent groups with maximum freedom of action and initiative, free from rigid regulations and prescribed channels, and were highly politicized. For decades, part of the ingrained culture has been to resist, defy, test, protest against and challenge official authority. It would be naïve to expect that this ingrained culture would disappear overnight.[26] In MK and APLA there was also little rank differentiation and few special privileges for anybody. This is in sharp contrast to the fifteen-level rank structure of the other statutory forces and the new SANDF. According to the Setai Commission, it has been far more difficult for former MK and APLA members to adapt as a result of these differences.[27]

This has created a clash in leadership styles, as former MK and APLA leaders have been promoted to senior leadership positions in recognition of the contribution they made to the struggle.[28] Although the former SADF (the largest of the integrated forces) still represents over 52.5 per cent of senior leadership,[29] it is of interest to note that presently at the rank of brigadier general and above, the former MK (30.8 per cent) and APLA (9.2 per cent) together represent 40 per cent of the top leadership of the SANDF, and the former TBVC forces 7.4 per cent (see Table 6.1). In this regard, the Democratic Alliance recently warned that the 'ANC is slowly but surely taking control over the defence force'.[30]

According to Perlmutter and Bennett,[31] one of the main dilemmas facing countries having to integrate revolutionary soldiers in post-revolutionary times into a professional military is 'how to disarm these soldiers politically and re-arm them professionally'.[32] This is one of the key challenges facing South Africa with the integration of former revolutionary forces into the SANDF and underlies much of the current racial tensions stemming from differences in leadership styles and past political loyalties.

Affirmative action and racism

in terms of the racial profile of the SANDF, the integration of the former revolutionary and TBVC defence forces have contributed largely towards normalizing the racial profile of the SANDF. The recent Human Resource Strategy 2010 stipulates that the force composition should be Africans 64.5 per cent, Coloureds 10.2 per cent, Indians 0.75 per cent and Whites 25.4 per cent. At present the overall composition of the SANDF as at 15 November 2005 just about conforms to these quotas, with Africans making up 65.1 per cent of the total force, Coloureds 11.3 per cent, Indians 1.4 per cent and Whites 22.1 per cent. However, this racial profile is not reflected at all in the different rank groups and professional branches in the SANDF.

At the operational level (middle management), Whites still make up more than half of officer and non-commissioned ranks. To rectify this, Blacks are being fast-tracked through the ranks by giving them preferential access to courses required for promotion and appointment to higher ranks. According to Honey, members of the former non-statutory forces have benefited greatly from affirmative action, and many loyal guerrilla fighters have been rewarded with high ranks, often without regard to experience, skills or age.[33] This has given rise to considerable tension, as Whites perceive standards to have declined for the sake of political expediency. A White army officer in the combat services expressed his concern with the challenge of representivity for operational effectiveness:

> Leaders that are placed and promoted should be competent. This is not the case, it is rather empowerment to please politicians and therefore members are forced into positions that they are not competent for. Hence, the SANDF is losing its competency.[34]

Similarly, top SAAF instructors and former fighter pilots say that the relaxation of standards, where pupils who only manage a 40 per cent pass rate still qualify as fighter pilots, is 'nothing short of a death sentence'.[35] A major challenge, given South Africa's legacy of apartheid education, is the inability to recruit and retain suitably skilled personnel. The Air Force and Navy face serious challenges as pilots, sea-going and engineering personnel, and highly skilled technicians leave and cannot be replaced with persons of colour. This has been aggravated by affirmative action policies that have driven too many Whites out of the armed forces and sidelined others' expertise and experience.[36] A comment by a senior White naval officer highlights the sense of alienation experienced:

> I feel that as a White senior officer with *scarcely* needed skills in a branch that is not attractive to Africans my service is no longer wanted or needed based on the colour of my skin. I find this very hard to accept as I serve my country and service just as loyally as in the past. This is expected of a military officer. My goal of attaining rank after a successful career is no longer

84   L. Heinecken

achievable. For the first time I need to think of another career in the medium term.

Conversely, Black officers feel that segregation is still rife and that there is still no true representativity. A Black major expressed the view that

[t]here is a lot of institutional discrimination perpetuated by individuals in senior positions against subordinates, practised predominantly by Whites against Blacks. This form of racism should be rooted out in the entire DOD.

As one witness in the Setai Commission stated, 'If you have black and white, you will have racism and if you have black and black of different cultures, you'll still have problems.'[37] From this it is apparent that negative perceptions about racism and discrimination in the DOD stem from various factors. This is exacerbated by the fact that commanders and seniors are reluctant to take disciplinary action for fear of reprisal or being accused of being racist or biased.[38] As the SANDF is still polarized in terms of race and former force, this makes the enforcement of discipline all the more tenuous. Allegations of racial inequality and arrogance, and counter-allegations of defiance and disobedience, all contribute to the reasons why command elements are reluctant to take action against members who are ill-disciplined. Former loyalties persist, and according to the Setai Commission it is a known fact that 'former revolutionary soldiers also generally refuse to testify against their comrades'.[39]

Another issue that should not be ignored is the impact of HIV/Aids, a disease that to date has disproportionately affected Africans, affecting not only race relations, but also absenteeism, discipline and morale.[40] Just for the period November 2004 to 31 October 2005, 845 deaths were reported – most in the age category 24–44 years. Although not all these deaths are attributable to Aids, one can safely say that Aids and affirmative action are rapidly eroding the institutional memory[41] and capacity of the SANDF.

Cultural practices and language

Besides this, there is the accusation that the SANDF is still not Africanized. Many of the present traditions stem from the former SADF and are Eurocentric (or, more specifically, British) in origin. Some Africans feel that little recognition is given to their cultural practices and beliefs, including the acceptance of customary marriages and dependants, weddings, funerals, godparents and the need to communicate with the ancestors.[42] Typical problems experienced include the lack of recognition by superiors of the importance assigned to certain cultural ceremonies, the need to attend funerals of extended family members and the desire to consult traditional healers.

In terms of policy, steps have been taken to accommodate certain cultural and religious practices. In accordance with SANDF dress policy prescripts, the following religious adornments may be worn, for example: the Zulu custom of

Isiphandla (a piece of cattle or goat skin worn on the wrist), Lakshimi string (a band of seven strands of red cotton), the cultivation of a beard and a moustache by men (to show Islamic, Jewish or Shembe orthodoxy), the fez or turban worn by men (prescribed by various Islamic traditions) and the official mourning button or band.[43] In addition, a special leave dispensation has been introduced that allows members up to five days' special responsibility leave to attend funerals.[44]

The African practice of having multiple spouses (up to five wives) has been accommodated, as well as customary marriages. Life partners are also recognized, whether these be of the opposite or the same gender. This entitles the dependents to enjoy the same benefits as dependants of Western marriages, including medical benefits. At present, the right to consult traditional healers instead of registered medical practitioners is not recognized and members have to take normal leave instead of sick leave if they wish to do so. This remains a point of contention, but increasingly, as time progresses, policies and practices are being adapted to accommodate African traditions and cultural practices as far as it is possible within the military context. Then again, many young Africans, who have come to adopt a Westernized value system, find themselves in conflict with their own traditional culture as they are pressurized to conform to traditional rituals such as circumcision, respect for elders, the role of traditional healers, and so forth.

Another aspect that has led to a clash in cultures is the tension between individualism and collectivism. On average, Whites display a strong sense of individualism, where merit and individual performance serve as strong incentives to achieve. Africans are more collectivist, favour working in groups and attach a high value to group conformity. Being an over-achiever, especially if this makes the group look bad, is frowned upon. Particularly on military courses, this has become a source of frustration for White members, who prefer to be evaluated on their own merits, rather than as part of a syndicate or group. In terms of group behaviour, while Whites seldom stand together, the principle of unity is strength is well entrenched in African culture.

Besides these cultural differences, language continues to be a divisive issue. Few Whites speak any of the different African languages, and Afrikaans, even eleven years after the formation of the SANDF, is still the lingua franca of the Army, the largest arm of service. Just as many Africans who do not speak Afrikaans feel alienated by its continued use, so do Whites, now in the minority, when Africans switch to a black language in the company of other Africans. Language is an important source of power, and although English is the official language of the SANDF, it is generally accepted that the dominant language of a region (for example, Zulu in KwaZulu-Natal, Xhosa in the Eastern Cape and Afrikaans in the Western Cape) may be used as a medium for communication.

Gender equality

Understandably, race has dominated the transformation process in the SANDF, and hence gender and the accommodation of homosexual rights have been less controversial, although not without their challenges. The constitutional provisions that forbid discrimination on the grounds of race and gender have meant that the SANDF was compelled to issue policy guidelines allowing women to be trained and employed in all ranks and positions, including combat roles. The Department of Defence equal opportunities and affirmative action policy stipulates that barriers that limit the absorption of women in certain branches be identified and eliminated.

Although most women still serve in traditional 'female' roles, since 1996 there have been women crews in the Infantry, Artillery and the Armoured Corps of the Army, as well as in the SA Air Force as pilots and in the SA Navy on ships. For example, in the SA Army 3.1 per cent of women serve in the SA Army Infantry Corps, 10.5 per cent in the Artillery Formation, 13.1 per cent in the Air Defence Artillery, 7.7 per cent in the Armour Corps, 6.7 per cent in Engineering and 16 per cent in Intelligence.[45] Women receive the same training as men and there is no gender discrimination except with respect to facilities and with certain drill adaptations that are considered fair discrimination based on physiological differences. The percentage of women in uniform has increased from an average of 9 per cent in 1995 (Table 6.2) to 17 per cent by November 2005.

In recent years there has been considerable political pressure to accommodate women in senior leadership and formerly male-dominated combat support and support branches. A survey conducted by the Inspector-General's Centre for Effect Analysis in 2001 found that almost two-thirds of the respondents agreed that it is possible to integrate women into special combat roles without compromising cohesion. However, a number of problems were raised concerning the deployment of women in front-line positions. Typical concerns revolved around the deployment of both parents, or single-mother families, the length of deployments, and the safety of women if they become prisoners of war. Although respondents felt that women adapt equally well in culturally diverse environments, most felt that the dominant societal culture, where women are still considered subordinate to the authority of men, poses a problem in an operational situation, and more needs to be done to sensitize people on gender equality.[46]

For women serving in the SANDF, the real issue is not whether they may

Table 6.2 Composition of the South African National Defence Force by race and gender at 15 November 2005 (%)

	Men	Women
Indians	1.2	0.3
Africans	55.2	9.9
Coloured	9.4	1.9
Whites	16.8	5.3
Total	82.6	17.4

Source: Department of Defence personnel statistics.

serve in combat, but whether they do so in a gender-friendly environment. Legally and politically, the mechanisms are in place to ensure that policies are implemented, but true equality cannot be achieved where the support structures are absent and attitudes that render women inferior and subordinate remain unchanged. For African women this can be directly related to the cultural subordination they experience within society, the greater childcare responsibilities they carry and the fact that they have been disproportionately the victims of violence and rape in South Africa. This prompted the Department of Defence to implement a defence instruction on the prevention and eradication of all forms of gender-based violence in the DOD. The policy addresses a wide range of gender-based violence, from domestic violence to sexual violence, violence against women during armed conflict, and violence in the work environment. This policy specifically states that the 'DOD will not condone any action by any of its members on the basis of custom, tradition or religious consideration that justifies gender-based violence'.[47] These issues raise questions not only for leadership and deployability, but for women's personal security when serving in isolated environments and when returning home after late-night duties. Women, for example, complain that they do not feel safe when performing guard duties alone at night.[48]

This is besides the realities of African deployment, where women have been the target of abuse, are often at a disadvantage in sexual negotiations and are more vulnerable to HIV/Aids infection. Of concern is that despite the noble efforts of the DOD to eliminate gender-based violence and prevent HIV prevalence within the ranks, the United Nations recently reported that the South African peacekeepers in the Democratic Republic of Congo (DRC) were responsible for a third of the sexual misconduct reported. These figures did not include seven other South African soldiers who allegedly committed crimes of sexual misconduct, or reports of similar incidents in Burundi.[49] With an increasing number of women being deployed on peacekeeping missions in Africa, immense human resource challenges lie ahead in implementing gender equality in the SANDF, challenges that stretch way beyond the politics of gender integration associated with women in combat.[50]

In terms of the rights of homosexuals in the SANDF, the DOD White Paper on Defence declared that it will operate strictly within the parameters of the Constitution and will not discriminate against anyone in terms of their sexual orientation. Any member indulging in or condoning unfair discrimination on grounds of homophobia and/or heterosexism is guilty of an offence. To assist in the breaking down of stigma, awareness programmes are conducted among the services and divisions to eliminate existing prejudices, and a policy is currently being formulated to combat discrimination in the DOD on the basis of sexual orientation.[51]

Although, legally, homosexuality is permitted, negative sentiments still prevail and it remains a 'silent right'.[52] A recent survey showed that substantial prejudice and stereotypes with regard to gays and lesbians exist. These range from concerns about the impact homosexuals may have on the spread of

HIV/Aids, possible abuse of authority (seniors harassing juniors), the impact that 'openly gay' personnel may have on morale, the impact on combat effectiveness, the sharing of the same facilities and issues of immorality. Most felt that homosexuals are still 'in the closet' because they are afraid of being rejected, scorned, oppressed, victimized, criticized and publicly judged. To address these negative sentiments, it has been recommended that more needs to be done to remove stereotypes and misperceptions about homosexuality in the SANDF through awareness campaigns.[53]

Management of diversity programmes

The SANDF is acutely aware of these underlying tensions. Ever since the formation of the SANDF in 1994, numerous programmes have been instituted to help reduce interpersonal conflict, increase employee knowledge on multicultural issues and create a shared organizational vision that could bridge interpersonal conflicts and promote racial tolerance. Most of the programmes try to instil a practice of mutual respect and encourage cross-cultural exchanges, whereby members can recognize and acknowledge different behaviour, mannerisms and demands flowing from a particular culture.

The Psychological Integration Programme

One of the first management of diversity programmes introduced after integration was the 'truth and reconciliation' process referred to as the Psychological Integration Programme (PIP). The aim of the PIP was for members to discuss their feelings of guilt, bitterness, fear or anger – to get these feelings 'out in the open' so that all could be sensitized to the different perceptions that exist. During this phase a video of apartheid was shown to highlight how the various groups experienced this period. The next phase addressed aspects of cultural diversity. The aim of this section was to cultivate respect for the backgrounds and cultures of others, with the hope of fostering a mutually acceptable military culture. The last phase of the programme dealt with cohesion and how to bring about a sense of unity among the diverse groups to be integrated. At the time, the intention was that everyone in the new SANDF would be 'PIPed'.

In substance the PIP was a worthy programme, but failed owing to the lack of adequately trained facilitators. Even though the facilitators succeeded in evoking emotions during the first debriefing phase, some lacked the group therapeutic skills to defuse the feelings of anger and mistrust. This proved extremely divisive, exacerbating tensions and making it virtually impossible to change the belief systems of members during the next cultural diversity phase, or foster cohesion in the final stage. Many White former SADF members were also angered by the emphasis placed on apartheid. Eventually, after about three years, senior army officers proclaimed that 'ons het nou genoeg gepip' ('we have now had enough PIP').[54]

With the unpopularity of PIP, some units adopted a different approach to the management of diversity.[55] For example, SAS Saldanha Naval Training base introduced it as part of its life skills course. Based on Albert Ellis's rational-emotive therapy (RET), known as the ABC theory of personality, an attempt was made to bring about change in the behaviour and attitudes of participants.[56] The approach was that 'it is OK to be what you are', but it is also 'OK for someone else from a different culture to be what they are'. If this is so, then one must learn to respect others and to address one's own prejudices and stereotypes if one is to develop a culture of tolerance. This coincided with Human's experience in diversity training in the private sector in South Africa, that the management of diversity is about how 'I see myself in relation to others and the value judgement placed on their behaviours and cultures'.[57]

Civic education and cultural diversity

With the end of PIP, management of diversity was included as part of the Civic Education Programme in the Department of Defence. The purpose of civic education was to instil respect among DOD personnel for the values of a democratic South Africa. Furthermore, the programme intended to make a significant contribution towards the building of cohesion, loyalty and discipline around a common set of values. In this programme the constitutional and legislative provisions pertaining to discrimination on grounds of race, gender, language and religion were discussed. The last phase of this course addressed the need to build a culture of equality and democracy within the SANDF. As part of the Civic Education Programme, this course was included in the curriculum of all the various military developmental courses.

As regards the section dealing with the management of diversity, the idea was that everyone in the DOD would gain an understanding of what management of diversity and equal opportunities entails. However, as with the PIP, the lack of adequately trained personnel to present the cultural diversity programme resulted in its being disbanded towards the end of 2003. Currently, moves are under way to replace this component of the Civic Education Programme with a section focusing on shared values. The DOD is due to implement a policy instruction on shared values,[58] which shifts the emphasis away from diversity to common values, with the hope of developing an organizational culture that is in line with the national values defined in the Constitution, as well as the strategic management principles of the DOD.

According to this policy, the core values that each member of the DOD should strive to internalize include professionalism, respect for human dignity, integrity, leadership, loyalty and accountability. The value most closely linked to the management of diversity is that of respect for human dignity, which emphasizes respect for others, tolerance of differences, the need to abstain from condemning and judging others, fairness in the treatment of others and effective communication to promote common understanding. Given the challenges of having to integrate so many diverse racial and military cultures, for the first time an active attempt is being made to create one universal military culture acceptable to everyone serving in the SANDF, irrespective of race, military background, creed, religion or gender.

Courses and seminars

In terms of the overall management of diversity, EO and AA in the DOD, the responsibility resides with the Equal Opportunities Chief Directorate (EOCD). Using as a basis the Equal Opportunities Advisors Course presented at the Defense Equal Opportunity Management Institute (DEOMI) in Florida, the EOCD has developed its own course to train people to manage diversity and deal with labour relations matters. The five-week course is designed to alert designated officers[59] to the nature, origin and ramifications of discrimination and to increase their knowledge and understanding of EO and AA. The topics discussed during these courses include aspects relating to socialization, communication across cultures, power and discrimination, racism, sexism and religious discrimination, sexual harassment, conflict management and affirmative action. The idea is that the person in the unit who is designated as being responsible for management of diversity serves as an adviser to the officer commanding on such issues.

Besides this training course, middle management are encouraged to attend a four-day management of diversity seminar – one for middle-ranking officers (captains and majors) and another for warrant officers and senior non-commissioned officers. This four-day seminar includes a briefing on the legal aspects regulating affirmative action and equal opportunities in the DOD, an understanding of what constitutes the management of diversity, and various other aspects relating to discrimination, racism, sexism, sexual harassment and people with disabilities. This course is divided into various phases. In the first phase participants are requested to expand on how they perceive the concepts of race, gender, ethnicity, culture, attitudes, personal differences and socio-economic class. This is interfaced with various exercises to concretize the concepts. Here the facilitators explain that there are different layers of diversity based on (1) one's own personality traits; (2) internal dimensions of diversity as a result of socialization; (3) external influences such as income, personal habits, religion, educational background; and lastly (4) organizational factors, which include aspects such as seniority, division, work location, union affiliation and management status.

Hereafter, how these four layers of diversity filter one's own perception of the world and of others is examined via various exercises – for example, a self-examination of how socialization influences behaviour. This is followed by discussions and exercises on how stereotypes are formed, for example about women ('all women are weak and sentimental') or on race ('all Whites are racists' and 'all Blacks are lazy'). Here members are requested to explain how these perceptions and stereotypes based on their own socialization affect communication and relationships in the work environment, and consequently how they affect mission readiness and goal accomplishment.

Having seen how diversity filters can influence work relations, the impact of communication in a culturally diverse work setting is examined. Specific attention is paid to factors that shape prejudice based on race and gender, the different sources of power that various groups possess (such as education, language, military background) and how certain stereotypes and perceptions influence one's behaviour. Other aspects addressed include the impact of language and the importance of ensuring that all understand, or share the same meaning of, a message and also non-verbal communication, which can be perceived to be contradictory. Attention is also drawn to those cultural practices, values and beliefs that appear offensive to others.

The final phase of the seminar looks at the benefits of having a diverse workforce and the long-term advantages of managing diversity effectively. By means of practical examples it is illustrated how this improves the utilization of the organization's human capital, can reduce interpersonal conflict, enhance mutual respect, foster a shared organizational vision and commitment, stimulate innovation (as more ideas from different viewpoints lead to better solution to everyday problems) and, ultimately, how this improves productivity and organizational effectiveness.

A separate two-day seminar is also presented to officers commanding, which aims to sensitize senior management to diversity issues at unit level. In this seminar the legal framework governing EO and AA is discussed, as well as the responsibility of the officers commanding, the leadership philosophy to be adopted, the current EO climate and the role of the EO adviser in managing diversity issues. Ultimately all the programmes are directed at creating awareness and a change in the attitudes and perceptions of members towards those coming from diverse backgrounds.

Concluding remarks

There is a belief that it is easier to manage diversity in the armed forces because everyone works towards a common goal or mission, does things in the same way, wears the same uniform and speaks the same jargon. While this may be true for most West European armed forces, in South Africa the lack of a mutually acceptable military culture has failed to bridge the divide between race, as well as between those coming from different military backgrounds. More than a decade down the line, people still club together on both sides of the colour line, and the events of just the past five years show how these divisions continue to impact on discipline, morale and leadership in the SANDF.

It is well known that any shift in power from one racial or ethnic group to another creates tension – even more so where the change is linked to a shift in power from one enemy force to another within the same institution.[60] In the SANDF it appears that loyalty and 'struggle credentials' have become more important than individual competency, skills and experience.[61] This has made the management of diversity in the SANDF more complex and trying. It has exacerbated not only the rift between Black and White, but also the tensions between the former 'enemy forces'. However, it is not a clear Black/White issue.

Not only White but also former African and Coloured SADF and TBVC members feel alienated by the growing dominance of MK and APLA, but are obliged to confirm for the sake of expediency.

On whether the management of diversity programmes have been successful, one can say that at the very least they have created a climate of tolerance and awareness. A major weakness is the inability of the EOCD to implement its programmes and initiatives on a scale that would make a significant difference. This lack of capacity is one of the key elements undermining the success of the current management of diversity programmes. Thus, the exposure of members to these worthy programmes remains limited, if not lost, and a more concentrated effort is needed to subdue the simmering racial and political tensions that undermine unity in the SANDF. A formidable challenge still lies ahead in achieving the objective of 'unity and diversity', and only time will tell what the eventual impact may be on organizational effectiveness – or ultimately on civil–military relations in South Africa.

Notes

I wish to thank Colonel F.J. Labuschagne, SSO Development, Chief Directorate Equal Opportunities and Affirmative Action, for his assistance in terms of providing details on the management of diversity programmes presented in the SANDF.

1 L. Heinecken, 'Managing Diversity in an Unequal Society: The Challenges Facing the South African National Defence Force', in J. Soeters and J. van der Meulen (eds), *Managing Diversity in the Armed Forces*, Tilburg: Tilburg University Press, 1999, pp. 187–210. Also see D.J. Winslow, L. Heinecken and J. Soeters, 'Diversity in Armed Forces', in Giuseppe Caforio (ed.), *Handbook of the Sociology of the Military*, New York: Kluwer Academic/Plenum Publishers, 2003, pp. 299–310.
2 Madiba is the nickname of former President Nelson Mandela, who was able to unite South Africans across the colour barrier during his period of presidency.
3 J. Kane-Berman, *South African Survey 2003/2004*, Pretoria: Institute for Race Relations, 2004, pp. 10, 29.
4 Ibid., p. 29.
5 From 1948 until the early 1990s the National Party practised a policy of strict racial segregation known as *apartheid*. This policy was designed to guarantee political and social domination of the country's White minority over the Black population. This past legacy has left a profound imprint on South African society and continues to divide Black and White as the present ANC government tries to correct past imbalances through various affirmative action and Black empowerment initiatives.
6 Kane-Berman, *South African Survey*, p. 226.
7 Ibid., p. 158; J. Herbst, 'Mbeki's South Africa', *Foreign Affairs*, 84, 6, 2005, 95.
8 Herbst, 'Mbeki's South Africa', p. 96.
9 *Constitution of the Republic of South Africa*, 108 of 1996, pp. 7, 107.
10 Representativity refers to the racial and gender composition of the SANDF and to the fair integration and equitable representation of the constituent integrating forces at all ranks within the organization.
11 *White Paper on Human Resource Management in the Public Service*, December 1997, p. 14.
12 Department of Defence (DOD), *South African White Paper on Defence*, May 1996, p. 32.
13 Ibid., p. 36.
14 Department of Defence (DOD), *Annual Report 1996/97*, p. 40.

15 Department of Defence (DOD), *Policy on Equal Opportunities and Affirmative Action*, DOD instruction: Pol. and Plan no. 00001/1998, dated 1 October 2002.
16 Department of Defence (DOD), *Department of Defence Policy on the Prevention and Eradication of All Forms of Gender-Based Violence*, DOD instruction: SG no. 00015/2002, dated 1 September 2003.
17 South African National Defence Force Training Order (SANDFTO), *Introduction of Civic Education in the Department of Defence*, ch. 5: Cultural Diversity, 1997, pp. 5–11.
18 G. Mills and G. Wood, 'Ethnicity, Integration and the South African Armed Forces', *South African Defence Review*, 1993, no. 12, p. 27.
19 J. Cilliers, C. Schutte, L. Heinecken, I. Liebenberg and B. Sass, 'Public Attitudes Regarding Women in the Security Forces and Language Usage in the SANDF', *African Security Review*, 6, 3, 1996, p. 6.
20 DOD, *South African White Paper on Defence*, p. 111.
21 Figure for KZSPF not provided in latest statistics.
22 Setai B, Final Report, Ministerial Committee of Inquiry, www.pmg.org.za/docs/2001/appendices/011012Setail.htm (accessed 4 November 2005).
23 P. Dube and G. Gifford, 'Killing Spree Triggers Rethink on Integration', *Sunday Independent*, 19 September 1999, pp. 1–2.
24 IISS, 'The South African Military: Racism and Restructuring', *International Institute for Strategic Studies*, 6, 10, 2005, pp. 1–2.
25 Country Report on Human Rights Practices – 2002, www.state.gov/g/drl/hrrpt/2002/18227.htm (accessed 16 January 2006).
26 Ibid.
27 Setai B, Final Report, p. 8.
28 Ibid.
29 This is due to significant change, as many White SADF officers accept the employer-initiated packages presently offered.
30 J. van Eeden, 'ANC "neem stadig maar seker beheer oor weermag oor"', *Rapport*, 20, February 2005.
31 A. Perlmutter and P. Bennett, *The Political Influence of the Military: A Comparative Reader*, New Haven, CT: Yale University Press, 1980, p. 23.
32 According to Perlmutter and Bennett, ibid., p. 21, the political motivation of the revolutionary soldier is integrated with the revolutionary movement. Successful revolutionary soldiers are not necessarily trained to be professional soldiers but have a high status in society, given their role in bringing about a new order.
33 P. Honey, 'The Battle for Survival', *Financial Mail*, 24 October 2003, p. 29.
34 These comments were made by officers attending the Senior Command and Staff Course in a Military Employment Survey conducted in July 2004 by the author.
35 R. Makings, 'Lower Standards a Death Sentence', *Sunday Times*, 30 October 2004, p. 38.
36 Anon., 'RSA Armed Forces Face "Danger" as Regional Security Roles Expand, Budget Declines', *The Star*, 31 August 2002, p. 15.
37 Setai B, Final Report, p. 30.
38 L. Heinecken, M. Nel and J. Janse van Vuuren, 'Military Discipline: Where Are We Going Wrong?', *Strategic Review for Southern Africa*, 25, 1, 2003, 93.
39 Setai B, Final Report, p. 32.
40 For a detailed discussion on the impact of HIV/Aids on the SADF, see L. Heinecken, 'Facing a Merciless Enemy: HIV/AIDS and the South African Armed Forces', *Armed Forces and Society*, 29, 2, 2003, 281–300.
41 Institutional memory refers to the knowledge and experience people acquire through the years while serving in a particular organization.
42 Setai B, Final Report, p. 34.
43 Department of Defence (DOD), SANDF Dress Policy Instruction Amendment no. 5:

Wearing of Religious and Medical Adornments by SANDF Members in Uniform, *HR SUP CEN/CER/R/406/11/B*, dated 25 March 2002.
44 Department of Defence (DOD), Defence Act Personnel New Leave Dispensation, 1 January 2003.
45 B. Buthelezi, 'Achievements of Women in Our Army', *South African Soldier*, 11, 12, 2004, 39–41.
46 EOCD, *Focus Group Results Regarding Women in Deployment Conducted in the DOD, October 2004*, DA/EOCD/R/106/30/5, dated 29 August 2005.
47 Department of Defence, *Department of Defence Policy . . .*, pp. 1–3.
48 Setai B, Final Report, p. 28.
49 'SANDF told of troops' Sexual Misconduct in the DRC', *SABC News*, 13 October 2005, www.sabcnews.com/africa/central_africa/0,2172,114291,00.html (accessed 29 November 2005).
50 L. Heinecken, 'Affirming Gender Equality: The Challenges Facing the South African Armed Forces', *Current Sociology*, 50, 5, 2002, 715–728.
51 Department of Defence, *Policy on Equal Opportunities . . .*, p. 15.
52 L. Heinecken, 'The Silent Right: Homosexuality and the Military', *African Security Review*, 8, 5, 1999, 43–55.
53 Equal Opportunities Chief Directorate, *Focus Group Results Regarding Homosexuality in the DOD conducted in the Western Cape from 27 January–6 February 2004*, DS/EOCD/R/106/30/5, dated 24 March 2004.
54 Comments by Lt. Col. (Dr) G. van Dyk, Department of Industrial Psychology, Military Academy, and former member of the National Training Team that developed and implemented the PIP.
55 At SAS Saldanha, this initiative was driven by the unit chaplain, Urie van Zyl, who had a keen interest in management of diversity issues.
56 G. Corey, *Theory and Practice of Counselling in Psychotherapy*, 2nd edn, Pacific Grove, CA: Brooks/Cole, 1982, pp. 173–174.
57 L. Human, 'Diversity during Transformation', *Human Resource Management*, no. 3, 1996, p. 8.
58 Department of Defence (Policy and Planning Division), *Department of Defence Policy on Shared Values*, Department of Defence Instruction Policy and Plan no. 00075/2002 – DRAFT, dated September 2003.
59 Previously this was the task of the multi-tasked functionary, which served at the various general support bases. However, with the restructuring of the SANDF, units now nominate a specific person in the unit to deal with these issues, such as the adjutant or communications officers.
60 M.B. Khanyile, 'The Ethnic Factor in the SA National Defence Force: Recruitment Strategies Revisited', *Politeia*, 16, 1, 1997, 78–96.
61 Struggle credentials refer to the contribution members made in the fight against apartheid, mostly by former MK and APLA members.

7 Diversity in the Eritrean armed forces

Mussie Teclemichael Tessema

Introduction

Eritrea, the youngest sub-Saharan African nation, became a sovereign state in 1993 after a thirty-year war for national liberation, which has become known as the longest liberation struggle in Africa. The Eritrean people resisted the Ethiopian occupation by peaceful means until 1 September 1961. On that day the revolutionary vanguard, under the direction of the Eritrean Liberation Front (ELF), decided to begin the armed struggle, with martyr Hamed Idris Awate as its leader. They numbered twelve people, with only ten old Italian rifles. By 1965 the ELF had about 1,000 fighters in the field. In 1970 the Eritrean People's Liberation Front (EPLF) emerged as an intellectual left-wing group that split from the ELF after large numbers of Christian highlanders had joined its ranks. By 1976 the ELF and EPLF had a combined force of 20,000 troops. They were making significant advances in controlling the rural and less populated regions of the country. The Christian-dominated EPLF and the mainly Muslim ELF struggled among themselves in a contested civil war (1972–74 and 1981), which ended after the EPLF had crushed its mother organization, the ELF, in 1981. The EPLF succeeded in liberating the country in 1991.[1]

Given the historical background of Eritrea, this chapter analyses diversity issues in the Eritrean armed forces in the context of two phases: during the liberation war (1961–93) and after independence (1993–2004). This is because the legacy of the Eritrean liberation struggle has considerably affected the current Eritrean armed forces. For each phase, diversity will be dealt with in terms of ethnicity and gender. In particular, the integration of women in the armed forces has drawn much attention as a *cause célèbre* within the war for independence. Besides ethnicity and gender, religion will be touched upon in this chapter as well, albeit more briefly. After each of the two descriptive analyses, a number of general conclusions will be drawn. The chapter will be rounded off with an afterword that briefly assesses the present state of civil–military relations in Eritrea.

During the struggle for independence, 1961–93

Smoothing out ethnic diversity

There are nine ethnic groups in Eritrea: Tigrinya, Tigre, Saho, Hedareb, Kunama, Nara, Bilen, Rashaida and Afar. The total Eritrean population is estimated at about 4 million. Roughly 50 per cent of Eritreans are Tigrinya and about 31 per cent are Tigre. The other seven groups are much smaller. They share the following percentages: Saho 5.0, Afar 5.0, Hedareb 2.5, Bilen 2.1, Kunama 2.0, Nara 1.5, Rashaida 0.5.[2] This demographic composition gives Eritrea a clear multi-ethnic character.

Each of the ethnic groups speaks its own language. Hence, nine languages are spoken, although about three-quarters of the population speak Tigrinya and Tigre.[3] Tigrinya is spoken mainly in the highlands, while Arabic has entrenched itself along the coast and the Sudanese border. In other areas the dominant language depends on the dominant ethnic group, each of which has its own native tongue. For the time being, Tigrinya and Arab have been declared to be Eritrea's official languages. With regard to religion, the population of Eritrea is equally divided between Christians (40 per cent Orthodox, 5 per cent Roman Catholic and 5 per cent Protestant) and Muslims.[4] It is important to remember that all these divisions within Eritrean society are not as clear-cut as might be assumed, and in fact no neat fit between regions and particular communities exists.

One of the main factors that contributed to the success of the EPLF was its ability to mobilize the overall population. According to Selassie, 'The EPLF's plan involved integrating all nationals, regardless of gender, age and ethnic origin, into action to satisfy the demands of the ongoing struggle and of the people themselves. Efforts were then made to make participation more comprehensive.'[5] The EPLF was successful in recruiting volunteer fighters from the various groups in Eritrean society.[6] Like other leftist revolutionary movements, the EPLF had as a basic principle equality between ethnic groups, religions, and men and women. The EPLF was modelled in large part on a Maoist-style mobilizing party and guerrilla strategy. A significant element of the EPLF's success was based on linking social transformation with military liberation and ideology with organization. Members were incorporated and socialized into the front through a six-month course of political education and military training.[7] In Eritrea the total dedication of all members to the liberation movement erased all other identities of family, religion, clan, class and gender. The fighters were thrust into the totally different world of Marxist-inspired guerrillas who preached equality between men and women and challenged all their traditional beliefs.[8] What made the EPLF one of the most formidable liberation movements in contemporary history was its determination not to reflect its indigenous social base, but instead to displace it with an ideology of 'Eritreanness'.[9]

The ethnic origins of the EPLF's army correlated to some extent with the size of the nine ethnic groups that constitute the Eritrean population. As Table 7.1 shows, the majority of the EPLF's participants were Tigrinya (64 per cent) and

Table 7.1 Ethnic profile of the Eritrean People's Liberation Front's army, 1993

Ethnic groups	Overall population (%)	EPLF's army (%)
Tigrinya	50	64
Tigre	31	24
Nara	1.5	3.3
Saho	5.0	2.5
Kunama	2.0	2.2
Bilen	2.1	1.9
Hedareb	2.5	0.9
Rashaida	0.5	0.1
Afar	5.0	0.8

Source: UNICEF (1997) and Demobilization and Reintegration Office (1993).

Tigre (24 per cent). The remaining seven nationalities accounted for about 12 per cent. In terms of ethnic categories generally, the EPLF mirrored the heterogeneity of Eritrean society, with the Tigrinya people more strongly represented and Afar strongly under-represented as compared with the national average.[10]

The EPLF stressed secularism and nationalism, and mixing of fighters from different ethnic groups. There was a conscious effort to shape a leadership that reflected the different communities. Its top leadership was made up of Christian highlanders and east coast Muslims, but in the course of the war, and as a consequence of the disintegration of the ELF, a preponderance of highland Tigrinya fighters developed.[11] This was, in part, due to differential patterns of flight: lowland pastoralists and agro-pastoralists became refugees,[12] while highlanders stayed in Eritrea. Some ethnic groups, like the Afar, were almost under Ethiopian control, which in turn limited their participation in the war for liberation.

Managing religion

It is also interesting to note that innovations were made in religious practices to eliminate divisions but still keep ceremonies sacred. It was decided that during all sacred ceremonies attended by fighters, the time formerly used for prayer would be occupied by silence. This type of practice respected everyone's point of view, while creating a binding ritual for those who believed in liberation. For instance, traditionally in Eritrean society it is taboo to eat meat prepared by someone of a different religion, whereas in the EPLF such a notion was eliminated.

In addition, for the first time there were many cases of intermarriage between Muslims and Christians. Silkin,[13] a British anthropologist, described how the EPLF encouraged the fighters to marry, contrary to traditional custom. She claimed that the EPLF tried to undermine the practice of marrying within one's own group, thereby breaking down traditional barriers between people of different ethnic, religious and social groups, which was conducive to the whole process of building an Eritrean national consciousness. The fighters chose their own spouses and were encouraged to get to know each other intimately before marrying. The intention was that, through this, they would be able to test their compatibility, whether they fitted together in all ways.

The important idea was to unite the different ethnic groups and thus assist the national project of building a nation. The EPLF's marriage law represented a break with traditional marriage as an alliance between two families.[14] As Pool observed, 'There were many marriages between those of Christian and Muslim backgrounds: a rare phenomenon among wider Eritrean society.'[15]

Women warriors in the EPLF

As was hinted at above, the subject of gender in the EPLF has drawn a good deal of attention.[16] This is for good reasons: Eritrea's experiences in this respect differ from those of many other countries in which liberation movements waged guerrilla warfare. In comparison, the EPLF had a high percentage of women in the armed forces.[17] Women were first allowed to join the EPLF in 1973. The first entrants were a group of four university students, three from Addis Ababa and one from Asmara.[18] Women were encouraged to join from that time onwards and throughout the war. Women's participation increased as the EPLF replaced the more religion-prone ELF in the forefront of the war.

By 1979, women constituted 13 per cent of the fighters and 30 per cent of the EPLF as a whole. In 1990, 40 per cent of the total membership of the EPLF and 25 per cent of front-line combatants were women.[19] As of 1993, women made up 34 per cent of the EPLF. The EPLF had a higher percentage of women than any other liberation army in the world. They were among the few women in the world who really did play an exemplary role in attaining independence.[20] This renders the Eritrean case an interesting, if not unique, experience.

The success of the Eritrean revolution relied to a large extent on the spread of its ideology among a sufficiently large part of the population via a programme of political education. Within this ideological framework, certain forms of traditional lifestyles were judged to be backward and a modernist outlook was propagated. An important part of this modernist agenda advocated the equality of women and men by virtue of their joint participation in the revolution.[21]

To women fighters, the new existence offered an alternative approach to gender roles. Women were encouraged to assert themselves and actively learn the ideology of the liberation army. Because of various factors like age, isolation from family and a total break with earlier values, they were thoroughly socialized into the guerrilla movement.

There are many reasons why Eritrean women joined the liberation struggle, but the occupation as such certainly was an important one. When a country is occupied, there is often a high degree of consciousness in the population, and many people want to resist the occupation. The political aims of liberation movements that wage war against the occupying enemy appeal equally to both men and women. Women want to resist and mobilize in ways similar to men.

A quite different reason for joining the struggle was the wish to escape an arranged marriage.[22] According to Worku Zerai, who was one of the four female fighters who joined the EPLF in 1973, the majority of peasant girls marry before they are 15 years old.[23] Many girls are infibulated, and the wedding night can be a traumatic experience for them. For this reason, several young girls, only 10 to 14 years old, ran away from their homes and signed up as volunteers.

Legesse also states that some uneducated men would go as far as to write a *nashuda* against the wife.[24] This meant that she could never marry again and was tied to the man for life. An ex-wife in this position was not allowed ever to have her own house. Facing such a situation, the only way to become free was by joining the revolution. Until the mid-1970s, 95 per cent of Eritrean women were illiterate. Women were not considered to have thoughts or opinions of any value, so no effort was made to educate them. One traditional Eritrean proverb states, 'Just as there is no donkey with horns, so there is no woman with brains.' Such sentiments led to women being prevented from taking any political role or entering into discussions of village affairs.

The link between the revolution and women's liberation is easy to see. In Eritrea, women pointed to the needs of the liberation movement when their own wishes clashed with those of their husbands and with women's more traditional roles. In fact, women who could not conform in their personal life saw the revolution as an opportunity to get out of a difficult situation: 'The movement drew many rebel women into the revolution ... transforming rebellion into revolution.'[25] Many liberation movements have included women's rights and equality for men and women in their programmes for political change. The EPLF thus developed an explicit position in support of revolutionary women.[26]

Particularly since 1974, the EPLF has brought extensive changes to women in Eritrea. Previously, women had supported the liberation struggle by providing food, shelter and information, but with the emergence of the EPLF as the main independence group, women's participation became more direct. Women served in all capacities,[27] except for the top ranks of the leadership, as 'women's entry into central leadership positions' was not considered a priority.[28] The Eritrean women fighters participated in every activity of the armed struggle alongside their male colleagues.[29] The EPLF was concerned to overturn traditional divisions of labour in domestic work. Some of the roles traditionally considered the preserve of women (cooking, cleaning, as well as fuel, wood and water provision) became the duty of all combatants regardless of gender. The EPLF is said by nearly everyone to have moved beyond the tokenism of viewing women as merely the supporting chorus.[30]

Women and men lived communally with little or no privacy. They shared in all the domestic tasks and enjoyed a large measure of camaraderie. Eritrean women's contribution placed the Eritrean liberation movement in a unique place in the revolutionary history of Africa and the world.[31] It has also been observed that women joined in the hardest times of the liberation struggle, when the Ethiopian Army was five to eight times stronger than the EPLF.[32] Therefore, it has been mentioned as an outstanding feature of the Eritrean revolution that women participated heavily and in all aspects in the fighting force.

Nevertheless, no women fighters ever achieved the three highest officer ranks. The military hierarchy starts with command of a *gujelle* (ten persons), then *ganta* (three *gujelle* – thirty persons), followed by *hailee* (three *gantas* – ninety persons), *battolini* (three *hailee*), *brigade* (three *battolini*) and *kefle-serawit* (three *brigades*). No women advanced further than *hailee* commander.[33] Moreover, a schematic overview of women's participation within the EPLF towards the end of the struggle shows that a traditional division of labour to some degree also existed within the EPLF.[34] Hence, the process of allotting tasks to men and women within military units was not free from gender bias, despite all revolutionary ideology concerning gender equality. It is probable that women's ability to reach a high rank was curtailed by their generally lower levels of education and work experience relative to men.

It is interesting to note that marriage was not permitted between the EPLF fighters until 1977. A strict rule of celibacy was imposed. After 1977, however, marriage became legal and women fighters were allowed to marry their comrades of the trenches. They did it with a supreme contempt for all the prejudices that had previously ruled weddings in their society. After a brief honeymoon, husband and wife returned to the front, often not seeing each other again for several months. During the course of the war, many women married,[35] but when their maternity leave ended, they returned to the battlefield. A woman was not supposed to spend much of her time caring for her child if she had one. Children were the responsibility of everyone and would be looked after by both male and female fighters in the EPLF kindergarten. Many mothers would have preferred to stay with their child, but had to conform to the culture of putting the struggle first and leave their children behind.

Sexual harassment was not a serious problem, for there were strict rules and regulations. In addition, the 'criticism' and 'self-criticism' practices of the EPLF played a vital role in disciplining the Army and keeping sexual harassment to an absolute minimum. 'Criticism' and 'self-criticism' were the most important instruments used by the EPLF to make its policy effective. Every unit in the guerrilla army had a criticism and self-criticism session once a week. This practice was very propitious for women: it made their life easier. A man could be criticized for underestimating what a woman could do in combat. A woman could be criticized for going to the kitchen too frequently. And a leader could be criticized for not treating his male and female soldiers equally.[36]

As an interviewed army official noted, 'self-criticism' and 'criticism' were something that the fighters in the EPLF dreaded. If an issue was important to the leaders, the offender would be isolated and ostracized. To escape being criticized, the fighters preferred to conform, even if they did not agree. Because of this way of sanctioning unwanted behaviour, the ideology of the EPLF had absolute respect. Since the goal of equality is crucial for the unity of a revolutionary movement, women were able to appeal to higher levels of authority if their rights were violated. In sum, the lives of women were made easier thanks to the criticism sessions.[37]

Last but not least, a new form of gender-neutral greeting evolved in the

EPLF. In Eritrean culture, women kiss three times on the cheek and men shake hands – unless they know each other well, in which case they also kiss. But in the field the new form of greeting became the shoulder kiss whereby people rub and bump shoulders while shaking hands. This genderless greeting has eliminated the social uncertainty concerning whom to kiss and whom not to.

Conclusions

Having discussed the representativeness of the Eritrean ethnic groups and women in the EPLF, one may ask how effective the diversity management was. If we define human resources management as the process of attracting, motivating and retaining staff members in order to achieve individual, organizational and societal objectives, it can be said that during the war for independence, the EPLF was rather successful in attracting, motivating and retaining its volunteer fighters. In this regard, the most important contributing factor was the presence of committed and effective leadership consisting of the following elements: narrow power distance, presence of 'criticism' and 'self-criticism' sessions, giving priority to the national cause rather than to individual and group-related issues, and intensive political education.[38] Most of the aforementioned critical factors are said to have been inadequate or missing in the post-independence period, which, as we will discuss in the next section, has shown rather different trends with regard to ethnic and religious diversity, but especially with regard to the position of women, in Eritrean society as well as in the Eritrean armed forces.

After independence, 1993–2004

Demobilizing, remobilizing and ethnicity

At the end of the liberation war (1991) the EPLF consisted of 95,000 volunteer fighters. However, in December 1992 the government officially announced a phased programme to demobilize about 60 per cent. Accordingly, from 1993 to 1997 the government demobilized about 54,000 fighters, 13,500 of whom were women.[39] As of 1998, Eritrea counted about 47,000 regular soldiers.[40]

In 1994 a national military service programme was introduced. The programme involved all 'women and men' between the ages of 18 and 40 undergoing six months of military training and twelve months of unpaid work in various ministries.[41] Since the border war with Ethiopia (1998), however, the vast majority of the national service conscripts have been serving in the Ministry of Defence. That is, the national service members were drafted to the front and the former demobilized fighters were remobilized.

During the border war about 300,000 Eritreans were mobilized – that is, about 10 per cent of the entire population, or more than 50 per cent of the population of working age. The percentage is highest for men, as they form 80 per cent of the fighting force.[42] Between 1994 and 2005, under the national military

Table 7.2 Ethnic profile of the Eritrean armed forces, 2004

	Overall population (%)	Armed forces (%)
Tigrinya	50	58
Tigre	31	29
Nara	1.5	2.2
Saho	5.0	3.0
Kunama	2.0	2.1
Bilen	2.1	1.9
Hedareb	2.5	1.4
Rashaida	0.5	0.02
Afar	5.0	2.4

Source: Ministry of Defence, 2005.

service conscription programme, eighteen batches were trained, with an average personnel strength of 12,000.[43]

The introduction of national military service has somewhat increased the number of members of the smaller ethnic groups joining the Eritrean armed forces (see Table 7.2). It is also interesting to note that smaller ethnic groups are well represented in the ground forces but less so in the Air Force and Navy, which are generally more technological and hence demand higher educational qualifications. Besides, an unstated but formal procedure, which is still in effect, is the continued exclusion of members of the Eritrean armed forces who belong to the small ethnic groups from certain posts that require specialized skills (artillery, tank, medical staff, finance, etc.). This is not a deliberate action, but a consequence of those groups having lesser educational qualifications. Nevertheless, the trend is improving, as the educational background of the small ethnic groups improves.

Religious tensions

During the liberation struggle, religion was not raised as a important issue, which in turn had a positive impact on normalizing the existing differences between Muslims and Christians as well as among Christians of different dominations: Orthodox, Catholic and Protestant. After independence, however, Jehovah's Witnesses refused to take part in the 1993 independence referendum and do their national military service. This resulted in October 1994 in a government directive that Jehovah's Witnesses had no citizen's rights. They were denied government employment, government housing, business licences and passports.[44] Eritrea's 1,600 Jehovah's Witnesses were the first religious group to experience repression. Members of Jehovah's Witnesses still experience official harassment ranging from prolonged detention for refusing military service to the revocation of trading licences and dismissal from the civil service.

During the war with Ethiopia, many Eritrean soldiers embraced various forms of Protestantism, reportedly alarming government officials and leading to the banning of prayer meetings among members of the armed forces. Attendance at such meetings is punishable by imprisonment. Religious repression has been particularly severe in the armed forces.[45]

Recently, Protestant Evangelical and Pentecostal churches, or 'Pentes' as they are collectively known in Eritrea, have begun to experience difficulties. The Orthodox Church first called attention to the growth of what it considers 'heretical' newer denominations, and the loss, particularly of its younger members, to these denominations. Since the early 1990s the independent Protestant churches have been able to attract a proportion of Eritrean youth. After they joined the Army, however, many of them have continued to practise their faith in the Army. This made the government issue a new rule that banned all the independent Protestant churches in May 2002 and only allows the four officially recognized religions, namely Muslim, Orthodox, Catholic and Protestant.

Women in society: the pains of transition

As the EPLF became the government and Eritrea was officially declared an independent state, the time was right for laws that would officially recognize the equal status of women. New laws gave women the right to own land, whereas feudal marriage laws were banned, bride prices and dowries were restricted, female circumcision was made illegal, women were given the right to vote, and citizenship for women and children born out of wedlock was legalized. However, despite the laws, the transition process from front line to everyday life was not easy. In practice, men retained privileged access to education and employment, and control of economic resources, with the disparities being greater in rural areas than in cities.

A study conducted by Leggese indicates that the transition process was very difficult, but it was much tougher for women fighters.[46] The women had married comrades in the ranks, and when they came back, many families put pressure on the men to divorce their wives because they did not conform to the traditional society's idea of a proper wife – who is meek, soft-spoken and gentle. Given the depth of patriarchal structures in rural Eritrea, there is little doubt that, even with enlightened government policies, change will come slowly. As a consequence, female ex-fighters experience a lot of tension in their lives, finding themselves considered somewhere between, on the one hand, heroines, and on the other, unclean women.[47]

With the military demobilization (1993–97), many traditional gender roles, especially in the countryside, have resurfaced.[48] Women face many problems, as they lack resources, skills and work opportunities, and at the same time the domestic role comes to the fore again. The fact that women participated as equals in the liberation struggle has only temporarily overshadowed the fact that they are defined in Eritrea by a strongly patriarchal society. In many communities, traditional customs remain more important than progressive laws, and the family remains the most powerful institution.[49]

Out of some 13,500 women fighters demobilized since independence, more than half are reported to have divorced, which is really disturbing. The marriages did not survive the return to a normal civilian life, which leaves the women in a precarious position. To put it somewhat differently, when they returned to civilian life, society required them to be resocialized anew. The resocialization has been painful, and not all ex-fighter women managed to find a life acceptable to themselves. Additionally, many women who were unmarried when demobilized are now past the traditional marriage age. Because of that they face difficulties when returning to their communities, as their assertiveness is resented, and images of free sexuality attached to living within the Army, make it difficult for them to find a husband.

Almaz Daniel has a story similar to those of many girls who joined the guerrillas in their teens. Born in 1962, she joined up when she was 16, in 1978. She was married at 24 and had a baby girl in 1988, but divorced in 1993. According to Almaz,

> When I married, I was mixing everything – politics and emotions. I had absolutely no sexual education; it was a taboo subject at home. When I met a young man who liked to talk politics with me, and who shared the same ideas, I was convinced he was exactly what I needed. We were obsessed by the will to build an exemplary society. After independence, however, the men changed; they have become traditional again. In fact, this traditional male thinking has deep roots that go back many generations. When they went to the front, men were forced to accept the EPLF policy of equality between the sexes. When they came back to the cities after liberation, the government had other priorities; it did not concern itself with the emancipation of women and men fell back into the old way of thinking.[50]

During times of revolutionary warfare one can often see posters of women with a rifle in one hand and a baby in the other, symbolizing a kind of 'superwoman' who combines her traditional nurturing role with her new role as fighter.[51] But after the war is over, things revert to the pre-existing tradition. That is, various revolutionary movements that mobilize women do not live up to their promise once the revolutionary leadership gains power.[52] Such posters have great relevance to the Eritrean case.

The EPLF's relative successes in incorporating women as equals relied more on repressing gender differences than on transforming gender relations. Hence, the objective of achieving gender equality often failed because the approaches followed were top-down and male-led.

Women in the military: less and less

After Eritrea gained its independence the government transferred thousands of women fighters into the public sector and demobilized about 13,500 women fighters.[53] As a result, the number of women in the Eritrean armed forces decreased substantially. After the introduction of military conscription in 1994,

Table 7.3 Representation of women in the Eritrean armed forces (%)

Gender	EPLF, 1993	NMS recruits	EAF, 1999	EAF, 2004
Women	34	40	20	10
Men	66	60	80	90

Source: Ministry of Defence.

Note
EPLF = Eritrean People's Liberation Front; NMS = National Military Service; EAF = Eritrean Armed Forces.

however, their number increased. When the 1998–2000 border war with Ethiopia erupted, women as well as men were mobilized within the national service framework and were fighting together on the different front lines. These *Warsai* women – as the new generation of women soldiers were called – comprised around 40 per cent of the national service recruits.[54] As of 1999, overall, women constituted 20 per cent of the Eritrean armed forces.

Although the number of women in the Eritrean armed forces increased during the border war (1998), women remain largely under-represented in the Eritrean armed forces. As of 2004, female personnel represented about 10 per cent of total personnel. From January 2000, with the exception of former EPLF fighters who had stayed in the Army after liberation, and some *Warsai* women in support staff roles, all women were recalled from the front line to participate in a wide-ranging skills development programme. The official rationale behind this move was to use human capacity in the most efficient way. Many observers, however, remarked that the main reason for substantially reducing the number of women in the Eritrean armed forces was the discontent of the majority of the Eritrean population.[55] This reaction influenced the government decision to shift the majority of women from national service in the Eritrean armed forces to other services in public-sector organizations. 'The Government, after a failed experimentation, has withdrawn almost all women from the frontline.'[56]

The general point here is that the integration of women in the Eritrean armed forces has not proceeded as before liberation. The presence of women soldiers throughout the military is uneven in terms of rank and function. Although officially there is no restriction, at present they are under-represented in the combat army. Women will be found disproportionately in the lower ranks. Given its basic male-orientedness, it is not surprising that the military has served to marginalize women from higher ranks. The consequence of this situation has been the reinforcement of women's general marginality in the armed forces in particular and other organizations in general.

With regard to sexual harassment in the Army, reports on sexual abuse are beginning to surface. Unlike the female freedom fighters of the pre-independence period, many young Eritrean women who enlisted in the national service have been experiencing sexual abuse. The practice of 'criticism' and 'self-criticism', which played a crucial role in minimizing sexual harassment during the liberation struggle, has been discontinued. Patrick Gilkes, a consultant on the Horn of Africa based in the United Kingdom, states that

> there is no doubt whatsoever that there have been cases of rape and sexual abuse at the 'Sawa' national service training camp, and while these have certainly been talked up by the opposition, the evidence suggests the numbers are considerable ... I think I am right in saying that the figure for HIV/AIDS in the general population is around 3–4 per cent, whereas it is 22 per cent in the armed forces.[57]

Many young Eritrean women have been forced to bear children by unknown fathers as a result of sexual crimes that took place in the 'Sawa' military training camps and on the front lines.[58] A study dealing with sexual abuse of women in Eritrea's national service, for which a sample of women of various educational backgrounds doing national service were interviewed, found more than 80 per cent of this sample had experienced some sort of sexual discrimination, ranging from verbal advances by higher officers to rape.[59]

Recently, the conscription of women has met with resistance by the majority of the population.[60] Most Eritreans do not support national military service for women for the following reasons:

1. Women suffer additional deprivation since every year they serve diminishes their chance of having a family of their own.
2. The fate of those women who fought for the liberation has not encouraged most Eritreans, let alone young female Eritreans, to join the Army or do their military service. This is because, in Eritrea, female ex-fighters are largely married to ex-fighters. But most male ex-fighters are married to civilian women, while female ex-fighters are hardly ever married to civilian men. Only 3.2 per cent of the women fighters are married to a civilian, compared to 96.4 per cent of the men.[61] The main reason is that male fighters prefer civilian women and thus women ex-fighters find it hard to meet new marriage partners. This illustrates a considerable difference between female and male fighters and adds to the the general difficulties women experience on returning to their communities[62] – as described earlier.
3. Women are not particularly attracted by a military career because it is no longer very prestigious and it has become a more dangerous career choice, especially after the border war with Ethiopia.
4. Sexual harassment in the Eritrean armed forces has become much more prevalent than before.
5. The length of service is unpredictable, just as is the nature of women's assignments in the armed forces.

Taken together, all these reasons have set off averse feelings among the general public towards the conscription of women for military service.

Conclusions

In developing countries like Eritrea, there is a very controlled release of information in regard to smaller ethnic groups and women in the armed forces. Nevertheless, the limited information that we have suggests that ethnic representativeness has been developing relatively successfully, while the position of women has seriously deteriorated.

The Eritrean struggle for independence has been rightly hailed by many as a really remarkable one because of the high percentage of women combatants. Women constituted up to a third of the liberation forces and were equally deployed in all spheres of activities together with their male co-citizens. Nevertheless, recently the number of women in the Eritrean armed forces has been reduced substantially to about 10 per cent of the total force. This is mainly due to the discontent of the majority of the population with the treatment of women under the conscription system.

Given the nature of the military profession, one should not expect gender representation to be the same as ethnic representation. Nevertheless, the prevailing situation has not been encouraging women to join the Eritrean armed forces. Hence, the under-representation of women in the Eritrean armed forces will grow ever stronger if the military do not catch up with evolving demographic trends.

Afterword

The Eritrean case is different from the other cases included in this book. First, Eritrea is only a few years old as a full-fledged country. It went through a thirty-year war for national liberation with more than 100,000 armed combatants under the leadership of the EPLF. The EPLF as a revolutionary and liberation movement was successful in recruiting armed forces (volunteer fighters) from almost all ethnic groups, as well as in attracting relatively huge numbers of women. Second, after seven years of peace (1991–98) a border war erupted with Ethiopia, which in turn led to the mobilization of about 300,000 Eritreans to the war front.

At present the Eritrean armed forces consist of two categories of personnel, namely former liberation fighters (those who were not demobilized and those who were demobilized but remobilized in 1998 ('*Yikealo*'), and those who were recruited through national military service conscription ('*Warsai*').

The indefinite years of service, and the *Warsais*' feelings that their education was totally stunted and their most productive years wasted, have all been incrementally eroding the morale of the Army. The problem is multifaceted and complicated, involving economic, political and managerial factors. The existing situation therefore has led Eritreans to flee the Army in droves. It can be argued that the high desertion rate of the educated part of the Army comes at a time when sophisticated armaments (manned by well-trained soldiers) are determining the outcomes of war. The little professionalism that has developed in the

Army is being further eroded as loyalty, and not ability, is becoming the key factor for promotion.[63] Thousands of young men and women have left the country since 2000 in an effort to avoid conscription or after deserting from the Army.

From 1994 to 2000, Eritreans did not object to doing their national military service. Since 2000, however, national military service has become unpopular as never before. For instance, the first batch of national service recruits in May 1994 comprised 10,000 youth, mainly from Asmara. Accounts of their service were widely broadcast on radio and television, and made many young people register for the second batch. Eventually, 30,000 youth had registered, while Sawa (the military training centre) can only accommodate a maximum number of 20,000 at a time.[64] During the war, many people volunteered to defend their country. It is mainly since the aftermath of the 1998–2000 border war and the deteriorating political, economic and human rights situations that people have been reluctant to do their military service.[65]

In Eritrea today, to conscientiously object to war is taboo. Conscientious objectors are considered to be cowards and people who lack patriotism. There is no alternative to military service in the civilian sector. Neither is there is a known military criminal court of law or other equivalent court of law where one can be judged impartially. The right to conscientious objection is simply not granted, and its consequences, and that of desertion, are severe torture, long-time imprisonment and the death penalty. Since the border war of 1998–2000, the number of conscientious objectors within the Eritrean military has increased nonetheless. At present there are thousands of Eritreans who have objected to military service. They are forced to leave their country and live in exile.[66]

Notes

1 In recent years, guerrilla warfare has increasingly gained importance in Africa. In contrast to the European understanding of guerrilla forces, in which they are perceived as being totally different from regular state armies, in Africa this kind of warfare can be regarded as the normal form of armed conflict. Sometimes guerrilla forces, like the EPLF, have been transformed into large, more conventional armies. Other guerrilla forces have not developed beyond a stage of insurgency. See C. Clapham, *African Guerrillas*, Oxford: James Currey, 1998.
2 *Children and Women in Eritrea: 1996. An Update*, Asmara: UNICEF, 1997, p. 2.
3 With the exception of the Rashaida, who use Arabic, all have their own local language (usually the same term refers to both language and ethnicity).
4 See www.school.alingsas.se/nolhaga/erire.htm.
5 W. Selassie, 'The Changing Position of Eritrean Women: An Overview of Women's Participation in the EPLF', in M. Doornbos *et al.* (eds), *Beyond Conflict in the Horn: The Prospects for Peace, Recovery and Development in Ethiopia, Somalia, Eritrea and Sudan*, The Hague: ISS, 1992, p. 68.
6 A. Mehreteab, *Wake Up, Hanna! Reintegration and Reconstruction Challenges for Post-war Eritrea*, Lawrenceville, NJ: Red Sea Press, 2005; D. Pool, *Eritrea: Towards Unity in Diversity*, London: Minority Rights Group, 1997.
7 Pool, *Eritrea*, pp. 12–13.
8 T. Woldegiorgis, 'The Challenge of Reintegrating Returnees and Ex-combatants', in

M. Doornbos and A. Tesfai (eds), *Post-conflict Eritrea: Prospects for Reconstruction and Development*, Lawrenceville, NJ: Red Sea Press, 1999, pp. 215–238; A. Wilson, *The Challenge Road: Women and the Eritrean Revolution*, London: Earthscan Publications, 1991, p. 132.
9. For details, see D. Pool, *From Guerrillas to Government: The Eritrean People's Liberation Front*, Eastern African Studies, London: Long House Publishing Service, 2001, p. xv; Clapham, *African Guerrillas*, p. 13.
10. Demobilization and Reintegration Office, *Survey of the Ex-combatants*, Asmara: Mitias, 1993; A. Mehreteab, *Veteran Combatants Do Not Fade Away: A Comparative Study on Two Demobilisation and Reintegration Exercises in Eritrea*, Bonn: BICC, 2002.
11. Pool, *Eritrea*, p. 12.
12. About half a million Eritreans lived in Sudan. See Mehreteab, *Wake Up, Hanna!*.
13. T. Silkin, 'Women Can Only Be Free When the Power of Kin Groups Is Smashed: New Marriage Laws and Social Change in the Liberated Zones of Eritrea', *International Journal of the Sociology of Law*, 17, 1989, 147–163.
14. The fighters' break with the norms and customs of civil society, however, backfires when peace finally comes. Marriages that were established contrary to the rules of civil society must pass a very difficult test in order to survive.
15. Pool, *Eritrea*, 1997, p. 12.
16. Wilson, *The Challenge Road*.
17. Mehreteab, *Wake Up Hanna!*, p. 156.
18. Wilson, *The Challenge Road*, p. 96.
19. Ibid.
20. See Pool, *From Guerrillas to Government*, 2001.
21. T. Muller, *The Making of Elite Women: Revolution and Nation Building in Eritrea*, Leiden: Koninklijke Brill NV, 2005, p. 2.
22. Silkin, 'Women Can Only Be Free ...', W. Zerai, *Participation of Women in the Eritrean National Liberation Struggle*, research paper, The Hague, 1994; A. Legesse, 'The Role of Women in Development', in *Poverty in an Area of Plenty: A Baseline Study for Development Planning in the Zula Plain of Eritrea*, Consultancy Report, Oslo: Norwegian Church Aid, 1994.
23. Zerai, *Participation of Women*.
24. Legesse, 'The Role of Women in Development'.
25. Wilson, *The Challenge Road*, p. 80.
26. Women have often played important roles in revolutionary socialist struggles, not only in Eritrea but also, for instance, in Nicaragua. While in other revolutions and liberation movements women have not been represented in such large numbers as in Eritrea, they nevertheless have been considered important for the outcome of the armed struggles. See Clapham, *African Guerrillas*; Wilson, *The Challenge Road*.
27. See also D. Connell, 'Strategies for Change: Women and Politics in Eritrea and South Africa', *Review of African Political Economy*, 76, 1998, 189–206; Zerai, *Participation of Women*; Mehreteab, *Wake Up, Hanna!*.
28. See A. Stefanos, 'African Women and Revolutionary Change: A Freirian and Feminist Perspective', in P. Freire (ed.), *Mentoring the Mentor: A critical dialogue with Paulo Freire*, New York: Peter Lang, 1997, p. 257.
29. See R. Pateman, *Eritrea: Even the Stones Are Burning*, Lawrenceville, NJ: Red Sea Press, 1990, p. 465.
30. Mehreteab, *Wake Up, Hanna!*; Pateman, *Eritrea*; Wilson, *The Challenge Road*.
31. G. Team, *Eritrean Women between EPLF/PFDJ Abuse and Opposition Neglect*, 3 March 2004, http://www.gabeel.com/Editorial/English/eng_editorial4.htm; *The Eritrean Profile*, 7 March 1998.
32. A. Veale, *From Child Soldier to Ex-fighter: Female Fighters, Demobilisation and Reintegration in Ethiopia*, Monograph 85, April 2003, Pretoria: Institute for Security Studies.

33 See A. Stefanos, 'Women and Education in Eritrea: A Historical and Contemporary Analysis', *Harvard Educational Review*, 67, 1997, 677.
34 For details, see *Children and Women in Eritrea*, p. 25; Selassie, 'The Changing Position', 69; Silkin, 'Women Can Only Be Free'.
35 More than half of them have children. See Mehreteab, *Wake Up, Hanna!*.
36 Zerai, *Participation of Women*, p. 22.
37 E. Barth, *Peace as Disappointment: The Reintegration of Female Soldiers in Post-conflict Societies: A Comparative Study from Africa*, Oslo: International Peace Research Institute (PRIO), 2002.
38 Zerai, *Participation of Women*; Wilson, *The Challenge Road*.
39 Mehreteab, *Veteran Combatants*, 2002.
40 See *Military Balance 1997/98*, London: Institute for Strategic Studies, 1997.
41 B. Weldegebriel and R. Iyob, 'Reconstruction and Development in Eritrea: An Overview', in Doornbos and Tesfai (eds), *Post-conflict Eritrea*, p. 36.
42 Mehreteab, *Veteran Combatants*, 2002.
43 *Statistical Data*, Beleza, Eritrea: Ministry of Defense, 2005.
44 *The Guardian*, 23 May 1995; *Country Reports on Human Rights Practices*, Washington, DC: US Department of State 1996, www.wri-irg.org/news/2005/eritrea-en.htm#Heading62.
45 For details, see Amnesty International, 2005, web.amnesty.org/library/Index/ENGAFR640172005; 'Country Report Eritrea', *Country Reports*.
46 Leggesse, 'The Role of Women in Development'.
47 Barth, *Peace as Disappointment*; Muller, *The Making of Elite Women*.
48 For more details, see V. Bernal, 'Equality to Die For? Women Guerrilla Fighters and Eritrea's Cultural Revolution', *POLAR*, 23, 2000, 61–76.
49 Connell, 'Strategies for Change'.
50 This story was told to the author by way of personal communication.
51 C. Enloe, *Does Khaki Become You? The Militarization of Women's Lives*, London: Pandora, 1988.
52 To mention a few: the Algerian revolution, the Tanzanian nationalistic movement and the Iranian revolution.
53 Woldegiorgis, 'The Challenge of Reintegrating Returnees'.
54 Muller, *The Making of Elite Women*.
55 Y. Gebrehiwet, *The Hollowing Out of the Army by DIA*, 2005, zete9.asmarino.com/index.php; Nebari-Hidri, *The Unsung Heroes of the Eritrean Struggle for National Liberation*, 2005, www.eritrea1.org/body.html.
56 Gebrehiwet, *The Hollowing Out*.
57 Available at www.ind.homeoffice.gov.uk/default.asp?PageId=4122.
58 Ibid.
59 For fear of repercussions, the author wants to remain anonymous. See Muller, *The Making of Elite Women*, p. 171.
60 Nebari-Hidri, *The Unsung Heroes*; Muller, *The Making of Elite Women*; M. Wrong, *I Didn't Do It for You*, London: HarperCollins, 2005.
61 Mehreteab, *Wake Up Hanna!*.
62 Muller, *The Making of Elite Women*, p. 171.
63 Gebrihiwet, *The Hollowing Out*.
64 *Eritrea: Frontlines of a Different Struggle*, Asmara: UNICEF, 1996, p. 20.
65 M. Tessema and J. Soeters, 'Practices and Challenges of Converting Former Fighters into Civil Servants and Their Management Afterwards: The Case of Eritrea', *Public Administration and Development*, 2006 (forthcoming).
66 Nebari-Hidri, *The Unsung Heroes*; Gebrehiwet, *The Hollowing Out*.

8 Diversity in the Indian armed forces

Leena Parmar

> Positive contribution of the Indian military is the consistent professional support they have given to Indian democracy.
>
> (Stephen Cohen, 1976)

Introduction

The Indian Army is the largest volunteer army in the world. India is the world's most populous society as well as the world's largest developing country with a democratic system. Surrounded by military-dominated and non-democratic states, it represents one of the most interesting cases of firm, stable and authoritative civilian control of the military. It is remarkable that the appalling problems of low economic development, sharp differences in income, mass poverty, over-population, illiteracy, ethnic antagonism, absence of any linguistic unity, cultural fragmentation, social diversities and a complex class system do not seem to hinder civilian control of the Indian military. The two neighbors of India – Pakistan and Bangladesh – are comparable or closely matched to her in a number of ways yet display divergent patterns of civil–military relations.[1]

India is a society of diversity, cultural pluralism and vastly different convictions. The Indian population consists of Hindus, Muslims, Christians, Sikhs, Buddhists, Jains, Parsees and other religious communities, speaking eighteen official languages as well as more than 544 dialects. If one travels in India for more than 200 kilometers in any direction, the language, food habits, custom and dress change all the time. Racial diversity can be observed from the North to the South and from the East to the West with all sorts of demographic differences. India has highly educated youth in urban centers along with illiterate and semi-literate tribal people and villagers. India includes simultaneously affluent people along with millions of poor who cannot even afford to have two proper meals a day. In a way, in present-day India centuries and epochs are all represented at the same time.

This enormous diversity in India has inspired many authors to reflect on the country's history, culture and identity. If we look at the two phrases – 'internal pluralism' and 'external receptivity' – from Nobel Prize winner Amartya Sen's recent collection of essays *The Argumentative Indian*, we realize how scientific temper has been the hallmark of Indian thought over the millennia. He points out how the twin features of 'internal pluralism' and 'external receptivity' have been woven into the development of Indian thought over the ages.[2] These twin features can be recognized when dealing with diversity issues in the Indian armed forces as well.

This chapter is divided into two main sections. In the first, 'Culture and history', I 'write around' the topic of diversity. I encapsulate it in a layered context of four interrelated themes: spirit of duty; caste of warriors; civilian control and military professionalism; and classical and contemporary missions. While I am dealing with these themes the subject of diversity more often than not is referred to in implicit and indirect ways. Yet in my view, in order for one to really begin to understand the case of India, this kind of contextual analysis is essential – not least because, as will be pointed out in the last paragraph of the second section, 'Modernization and socialization', the Indian military are still somewhat sensitive and secretive about some aspects of diversity. This is the reason why I will put forward a research agenda – diversity as a topic very much included – that would benefit the Indian armed forces as well as Indian society.

Before that, however, notwithstanding the scarcity of sources, in this second section I will directly address a range of diversity issues related to language, religion, community and gender. The chapter will conclude with some final remarks and recommendations on Indian civil–military relations in general.

Culture and history

Spirit of duty

In early times, India was a conglomeration of princely kingdoms constantly at war with each other. This resulted in their inability to unite against a common foe. The richness of India's resources and the absence of a united front attracted the attention of invaders from distant lands; a lack of unity and knowledge about the latest weapons and techniques predisposed defeat of these individual kingdoms. The invasions started with the Aryans, who were followed by the Persians and then the Greeks. There followed the Saka, Parthian and Kushan invasions, then the Huns, the Arabs, Mahmud of Ghazni, Muhammad Ghori, Timur, the Moghuls, Nadir Shah, Ahmed Shah Abdali, the Dutch, the French, the Portuguese and finally the British. Sadly, not only did rival kingdoms fight and destroy each other, but they also invited foreigners to help them against one another. Most of the invaders settled down in India, intermingled with the population, became assimilated, enriched its culture and became full-fledged Indians, readying themselves to meet the next aggressor. Thus, over a period of time, India became a multiracial, multi-religious, multilingual and multi-ethnic country, developing its own composite culture and ethos of unity in diversity.[3]

This many-sided Indian culture has its roots in the Vedas and other ancient scriptures, which still infuse their spirit in Indian society. The Indian classic

Gita, the divine song, symbolizes true Indian culture, particularly the ethics of soldiering. The first of the two most profound lessons from *Gita*, which stand out as the sterling ethics of Indian soldiery, is 'duty for duty's sake', which is further reinforced by a most sagacious piece of advice that Lord Krishna gave to the soldier Arjuna on the battlefield: 'The concern is with action alone and not with the result thereof.' When one does his *karma* (work/duty) as *pooja* (worship), there is no question of seeking appreciation or fearing condemnation. Work done as duty brings no frustration. From the above, it follows that the essence of duty is *nishkam karma* (desireless action). However, *nishkam karma* does not mean *karma tyaga* (renouncing action) or working aimlessly. The second eternal lesson from *Gita*, for a soldier, is 'duty unto death'. Lord Krishna encourages Arjuna by reiterating, 'Lucky are those soldiers who die for the righteous cause.' He exhorts, 'Slain, thou shall go to heaven; victorious, thou shall enjoy the earth. Therefore, arise, oh son of Kunti, resolved on the battlefield' (*Gita* II/37). Deep-seated in Indian philosophy is humility and magnanimity in strength, an essential part of the Indian character.[4]

These are characteristics of Indian society that have shaped the pattern of ethics among common people as much as they have influenced the ethics of the Indian military system. The characteristics of ancient civilization, tolerance, assimilation of different cultures at different points of time in history, various religions, various races and ethnic groups and sub-groups, along with concepts of *karma*, *dharma*, *sanskar* and *purushartha*, have made India a strong secular nation emphasizing the idea of unity in diversity. The constitution of India with its salient features has contributed to making India a robust nation. The secular aspect of Indian democracy has made her a very tolerant nation too. Although there are religious, language, ethnic, regional, geographical, racial, demographic, cultural, tribal and caste diversities, India has shown herself to be a remarkable nation with the notion of unity in diversity.

Caste of warriors

Prevalence of the caste system and its deep and pervasive influence on all aspects of life is a remarkable characteristic of Indian society and culture. Although stratification based on birth is common to most pre-modern societies, the intricate norms of marriage and sharing of food and water which were strictly imposed by the Indian caste system and formed an integral part of it are surely unique. Such complex rules governing various aspects of life are found in no other society. The caste system thus is a unique feature of Indian society.[5]

One particular element of this typical Indian constellation goes back to the ancient Indian social order, which was divided into four *varnas* (castes), i.e. Brahman, Kshatriyas, Vaisha and Sudra. Of these, it was the Kshatriyas who joined the armed forces. They not only constituted the state's soldiery, but also ruled the state. Thus, soldiering was deemed a noble profession, a call to one's supreme duty. The caste system in India, going back over a millennium before Christianity, also played a role in fostering professionalism within the military.

The system sanctified the role of various groups in society and enjoined on them a certain duty. It gave the warrior caste a vocation that was to last not merely for life but indeed for generations. Thus, the military life has always been seen as more than a career in India. It has always been considered to be a form of government service in a society where this was seen to provide stability, prestige and an assured income that agriculture was increasingly unable to provide. But a career in the Army has always been perceived as something more: a lifelong vocation, and even a career across generations. In the highly competitive recruitment to the ranks, parental service is considered a distinct advantage. Among the officer corps as well, the numbers of second- and third-generation officers grew during the 1990s after increases in pay and perks, and the introduction of better service conditions.

Lord Roberts, who served as commander-in-chief of the Indian Army from 1885 to 1893, can be credited with – or blamed for – introducing the policy of fostering the so-called martial races. Roberts's outlook was essentially that of an unsentimental professional, more interested in building up the Indian Army to meet the Russian threat than in keeping it fragmented in internal security duties. Roberts argued that the Russians, not the still docile Indians, would have to be faced, and the former could not be met with an army of inferiors. In his view the best available 'material' came from the northwest quarter of India (i.e. the Punjab area), and in his view the Army should be recruited from that particular area in the country. He also felt that the class regiment fought better, and the classes that were recruited should be thus organized. The British liked the 'yeoman peasant' who was independent, sturdy, upright, honest and reliable. In some cases landowners were acceptable, and some units drew from remote or nomadic tribal groups, but the favorite was the moderately wealthy farmer who had the right 'outlook.' Outlook and attitude towards soldiering were important criteria for recruitment. On the whole, the British played down the religious aspects of caste by stressing occupational and class characteristics of the prospective sepoys, non-officers or other ranks.[6]

Civilian control and military professionalism

The Indian military have been unique among the armed forces in the developing world in several ways.[7] They have never launched a military coup, for example, or attempted to usurp civilian power, not even when their relative pay, power, and influence in civil society declined sharply immediately after independence in 1947. The armed forces have remained scrupulously neutral over nearly sixty years of the country's independence. Even frequent calls to aid civil authorities in natural calamities have not distracted them from their primary professional responsibilities.[8] In a sense, India has had a First World military, even as the nation has remained part of the Third World. This is as much a matter of ethos, outlook and professionalism as it is a question of maintaining a strict attitude of neutrality, reasoning and distance from civil political affairs. Why has this been so?

For an answer, one must turn to Samuel Huntington's classic theory of military professionalism. Huntington argues that a professional military is 'politically sterile and neutral' – concentrating on military strategic matters and leaving political decisions (affairs of governance in today's words) to civilian authorities. This, he contends, allows the desirable condition of 'civilian control' over the military. Others have suggested that the term 'civilian supremacy' should be used to define this condition. In the Indian situation it has more often been 'civilian control' – and yet it has worked.[9]

The Indian Army has always been apolitical and, as a consequence, it epitomizes the heterogeneity of Indian society, which has led to positive unintended consequences for the role of the Army in the national integration. The Indian Army has grown and evolved through experiences and shared ethos of its tradition and glory. The shared tradition of the Army is characterized by its apolitical nature, which has revealed remarkable continuity despite changes in the type of government at the center. The country has witnessed several political crises, but all political analysts in the country have always ruled out the possibility of Army dictatorship. This is not only due to the democratic institutions of the country, but largely because its officials have always scrupulously avoided any alignment with individuals or political parties of the country.

Is the Army merely a profession like any other, or is it something more? The answer to this is 'tradition', which is the bedrock on which the Army is built. Army men see with admiration the famous lines of Field Marshal Chetwode in the hall of the Indian Military Academy, named after him, exhorting the young leaders, almost in communion, to rise to the occasion, because 'the safety, honor and welfare of your country comes first, always and every time; the honor, welfare and comfort of the men you command come next; your own ease, comfort and safety come last, always and every time'. The Army has grown over the years into its present healthy organizational state to meet the changing requirements of changing times. 'Tradition does not mean that you never do anything new, but that you never fall below the standard of courage and conduct handed down to you.'[10]

Classical and contemporary missions

To comprehend its ethos even better, it is important to understand the broad characteristics of the missions of the Indian Army, historical as well as present-day ones. The British Indian Army had a ninety-year history before India became independent. For another 100 years and more before that, it was the army of the British East India Company, with all the trappings of a regular army.[11]

The Indian Army fought during World Wars I and II and in many campaigns in Asia and Africa over the past two centuries. During World War I, India sent as many as one million soldiers to Europe to fight for a cause that was not hers. This was repeated during World War II, when more than double that number were sent overseas to Europe and Southeast Asia to fight for the British Empire.

Sociologically it is important to stress that this did not happen from any sense of loyalty to the throne of England or to the Viceroy of India. As a sense of nationalism was lacking, there was no notion of ensuring Indian security and enhancing national interests. The motivation was primarily allegiance to the regiment. It was the desire not to 'let the side down', a sentiment not easy to describe, that really encompassed the contours of professionalism. This was the basis of the Army's corporate identity and this remains so largely today.[12]

By the end of the World War II, Indian nationalism had developed on strong lines under the leadership of Mahatma Gandhi and the Indian National Congress. In addition to demands for independence, Indians fought for the inclusion of Indians as officers in the armed forces. Under pressure from nationalist leaders, the British government finally gave in and opened the doors to Indian officers in the armed forces. The division of the subcontinent and the horrors of partition marred the birth of India as an independent nation. The thrill of independence was thereafter influenced by a war with Pakistan over Jammu and Kashmir. Since then, war between the two countries has broken out on three more occasions: in 1965, in 1971 and in 1999.[13] As yet, the conflict over Kahsmir has not been settled once and for all. Efforts to do so by diplomatic means are serious but fragile.

Today, India remains firmly committed to the peacekeeping endeavors of the United Nations. India's contribution to UN peacekeeping operations since 1950 has crossed the 70,000-troop mark, which is the highest by any country in the world. The professional spirit of the Indian soldier is evident from the fact that for every new mission established by the United Nations, the first request to contribute troops is made by the United Nations to India. Since their first commitment in Korea in 1950, Indian troops have participated in some of the most difficult UN operations – particularly in Africa – and won universal admiration for their professional excellence. In support of UN peacekeeping endeavors, the Indian Army has contributed some outstanding force commanders, elite military contingents, impartial observers and dedicated staff officers. Their devotion to duty and excellent performance have been widely acclaimed.

The overall impression is that Indians working with UN missions in multi-ethnic societies are fair and impartial, because Indians themselves belong to a multicultural, multi-religious society based on a secular, democratic polity. India is a large and powerful country and it rightly seeks a permanent place in the Security Council. As India seeks such a responsibility, it should also be ready to meet more international challenges and obligations.

The effectiveness in worldwide peacekeping is parallelled by the performance in internal conflicts – an occasional role for which the military are far from eager, but when called upon will fulfill professionally. During riots, the moment that word goes around that the Army is going to appear on the scene, the murderous passions cool. Why else do riots – that police forces cannot control by firing bullets and tear gas shells – fizzle out at the very sight of the Army's olive green, even though the Army, under the Indian Constitution, is only called out 'in aid of civil power' and has to await a civil magistrate's orders before opening fire. In the words of an astute observer:

It is interesting to note that in India it is exactly the opposite with the police, because Indians consider the police much as a handmaiden of the local political establishment they usually hold in contempt. Never in the history of independent India has a unit of the army faced even a fraction of distrust from any quarter of the Indian population.[14]

Modernization and socialization

Organizational change

From a base strength of around 280,000 soldiers in 1947, the ceiling in personnel was raised (after the conflict with China in 1962) in 1964 to 825,000, the increase to be implemented over the following five years. Since mid-1980s this ceiling has gradually increased and today the Army's strength is one million plus.

Despite their traditional Army heritage, Indian soldiers reveal a high degree of modernization, which symbolizes the Army's dynamic flexibility in acceptance of innovations in ideas and action both in the external system of weaponry and in the internal systems of values and attitudes. Here again one recognizes Sen's earlier-mentioned twin features of India's 'internal pluralism' and 'external receptivity'. This leads to consonance and adaptability between the mechanical system of the Army and the personality system of the soldier. The dynamic flexibility of the Army is revealed in its acceptance and use of the latest technological systems of weaponry as well as the rational scientific attitude it shows in action and role performance. This is an objective indication of the process of modernization, which operates both at the level of organization and at the level of personality. The contemporary military organization has to be highly adaptive, as the skills and dimensions of war are becoming more and more lethal and dangerous. Unless the organization and the people who function in the organization are dynamic enough to accept the change, the system will be ineffective. Such a consonance between modernity of weaponry and modernization of ideas and thought is responsible for the effectiveness of the military system. This also signifies that the Army needs both, commitment and high morale on the part of soldiers as well as adequate technological sophisticated weaponry capable of matching the strength of the enemy. Battles and wars have been won and lost on both accounts – outdated weaponry as well as soldiers with low morale and commitment.

Handling heterogeneity: language, religion, community

The Indian soldier has a heterogeneous social and cultural background, which in fact is a replica of the Indian society and culture. The recruits come from varying regions, linguistic backgrounds, castes, religions, classes and family backgrounds, and hence they differ in their process of pre-Army socialization. It is significant to note that despite this heterogeneity of background, the military system has superimposed upon them the uniformity and homogeneity of the military culture of routine work, duty and lifestyles. Perhaps more than the militaries in other parts of the globe, the armed forces in India are successful in socializing recruits from all sorts of backgrounds into one military culture and way of life. This synthesis of micro-level distinctive background and macro-level military superimposition has created a degree of national integration that can hardly be seen among the civilian population of India. More and more people from castes that earlier were thought to be related to either the performance of sacred rituals or trading are nowadays entering the military forces.

The social backgrounds of the soldiers reveal differentiation and heterogeneity, yet the professional socialization of the Army makes them share similar subcultural and behavioral patterns, especially in relation to argot, dress, physical fitness and values of discipline, punctuality and loyalty. This may be illustrated by the way the Indian armed forces deal with language diversity. Although the recruits are from different language areas of the country, there is only one language in use in the Indian military, which is Hindi, in particular Roman Hindustani (where the script is in English). During the course of their training, all soldiers have to pass three examinations in this official language. Language training, hence, is a compulsory part of their total military training schedule. This also applies to the officers, who all have to master Hindi to be able to communicate with their troops. Some officers in addittion learn regional languages to have a better understanding of the troops.

To mold heterogeneous categories into one homogeneous community is the result of the Army tradition as well as rigorous and disciplined training. There is no other large-scale organization that has such a primordial goal of serving the nation and necessitating thereby the supreme sacrifice of life itself, should it be required. In fact, even the most primary institution, the family, does not ask for the sacrifice of life. It is only in the Army that members are prepared to sacrifice their life for the nation and its security. The Army reveals a high degree of organizational altruism, which is inculcated through a well-organized mechanism of socialization and training.

Indian soldiers' high degree of modernization is the result of their professional training and skills as well as the Indian Army ethos, which is universalistic, secular, rational, and oriented to the primordial goal of protection of the nation. However, no matter how universalistic this ethos may be, there is also room for diversity in the armed forces.

Religion plays a very active and important role in the Army. Places of worship have always been maintained in each unit. Only in the Indian Army it is possible to see, in the same unit, a temple, a mosque, a church and a gurdwara, all housed under tents separately, but in very close proximity to each other. Prayers are held in each tent, often on the same day, without any problems of friction. Festivals of each religion are jointly celebrated. It can be said that the military are still the only truly integrated secular organization in the country, paying respect to religious diversity nonetheless.

Also, in another connection one can recognize the way the Indian armed forces carefully try to balance unity and diversity. The various regiments of the armed forces reflect the diversity of castes, communities and ethnic groups. Some of the regiments are purely regionally based, like the Maratha, Madras and Dogra regiments, but there are certain regiments that have a mixed composition of troops. Even today, by and large the companies are constitued by various castes, communities and religious groups such as Meena (a tribe), Rajputs (a caste), Muslims (religion) and Gujarati (a community), and all such companies can be found in one single regiment. This way of structuring goes back to the British days and persists to the present time, which helps to give some sort of cohesion, motivation, competion and fighting spirit among the soldiers. Each company can practice its cultural activities and religious festivities, and can uphold the tradition of its caste and region. This in turn helps soldiers to strive for better integration in the armed forces. Officers do not come under caste and communal division; they are treated as a class apart. An officer is posted to any regiment, with no consideration as to region, community or religion. In these ways the Indian armed forces try to appease the centrifugal forces of human diversity in the organization. And they are successful at this.

The military organization, with its built-in stratified social structure, creates new modalities of leisure within the subculture of the Army. Officers and *jawans* (non-officers – other ranks) coming from different caste, class or religious backgrounds live together in a society of their own. Their disciplined life and hard working conditions are also reflected in their leisure activities, which are also disciplined, with a secular outlook. Consequently, they are bound together in a different human relationship with a positive approach towards life in general; and this promotes in them better cooperation, understanding and sympathy with each other. This is reflected during wartime as well as in peace, and also during their general duties. The type of leisure enjoyed by the military reflects the characteristics of their organization, which leads to a consolidated feeling of oneness. The military organization, with a built-in subsystem, has a unique pattern of subculture, which is responsible for a relationship that in turn is reflected in the work and lifestyle. The ceremonies, the medals, the flag, the parade, national anthem, the dress, the mess culture, etc. reveal the symbolic aspects of subculture, which enhance morale, self-image and the gratification dimension of personal and familial needs.[15]

Integrating women soldiers

The role of women has been increasing steadily in the field of India's national defense. Eligible women are recruited as officers on a short service commission basis in various branches of the Army, Navy and Air Force. In the Indian Army, women candidates can be inducted in the Indian Army against certain identified vacancies in various arms and services through the Women Special Entry Scheme (Officers). Women are offered short service commissions for a period of ten years, extendable by an additional four years. The annual intake is 150 with effect from September 2003. Presently, there are approximately 921 women officers serving in the Indian Army. The Indian Navy first recruited women in 1992, and a total of 179 women officers (including fifty-eight medical officers) are serving in various units in the Navy. These officers are assimilated into the mainstream, and their promotion prospects, training as well as career progression are on a par with those of their male counterparts during their tenure.

In the Indian Air Force (IAF), the recruitment of women as short service commission officers in flying, technical and non-technical branches also commenced in 1992. Presently there are about 515 women officers in the IAF. Although women officers are currently not being granted permanent commission, they can serve up to fifteen years in the IAF. The initial term of employment is ten years. Extension up to fifteen years is granted on a case-to-case basis depending on individual merit. The intake of female officers in the IAF has shown an upward trend during the past three years. In the years 2002, 2003 and 2004 respectively, sixty-eight, seventy-nine and eighty-three women officers were commissioned in the IAF.

The recruitment of female officers was based on populist considerations more than on military necessity. A recent study into the position of women officers by Colonel D.S. Randhawa demonstrated that, to the majority of the troops, the presence of women in the forces meant lowering of physical standards, adjustments of work culture and norms suitable to women, restrictions on a soldier's ego and freedom, tensions, courtship, jealousies, favoritism, disintegration of hierarchies, unenforceable codes of contact leading to resentment, and sex scandals. One of the questions in the study, related to operational performance, was phrased as follows: 'Are female officers willing to lead the male troops on a patrol or ambush duties?'

The findings were mixed. Younger women officers who had been serving for one to four years felt thrilled about this adventurous activity. Somewhat older, married women officers with five to eight years' working experience, however, considered this to be inappropriate, feeling nervous and bewildered at the thought of being a lone woman among male soldiers. Family, children and husband remained their major concern. Young male soldiers felt that their responsibility would increase in such a situation. Some senior male officers were evasive and non-committal, while the majority were not in favor of sending female soldiers on night duty or on patrols, ambush and convoy protection duties in counter-insurgency areas.

A few cases of adultery, stealing affection and illicit relations are noticed where both genders are equal partners. According to rough estimates, only 2 percent of cases are reported, and most of them are settled out of court, with the culprit apologizing for the criticized behavior. Military men need to accept the fact that women officers are here to stay, and more will join, and gender equality will occupy center stage in the twenty-first century. This will happen in the Indian armed forces just as it is doing nowadays in so many other armed forces across the the world.[16] The current proportion of 0.12 percent of women in the Indian armed forces is still very low, but the numbers will undoubtedly rise.

Hidden problems and research agenda

The military system necessitates a high degree of closure due to its role in the areas related to security and defense of the country. There is an in-built resistance to revealing facts about the internal dynamics of the military organization, because it is not in the nation's interest to reveal psychological and social information about soldiers; at least, that is the view of the Indian Defense Department. The Indian military organization has therefore not been studied widely by social scientists, because of the possibilities of misuse of the information by hostile forces.

There are institutional and structural ways to locate Army headquarters in an isolated or secluded area away from the civilian population, so that necessary physical, social and cultural distances are maintained. Army officials are required to adhere to certain well-laid down rules of conduct, and there are different mechanisms of rewards and punishment for them. Army information comes under the classified category, and it is rare for newspapers or journals to publish details of interpersonal, relational dimensions of soldiers. Nor do Army personnel make public statements on matters related to their service. These in-built aspects of protective cover provided for the Army make it difficult for social scientists to penetrate Army life and undertake any research, since all academic studies are required to be made public.

The sources for understanding the sociological dimensions of the military in India are largely autobiographical and biographical memoirs, records, reports and newspaper reporting. Such sources are certainly useful, but they have meaningful biases, revealing self-glorification or national patriotic overtones. An objective, facts-based approach to the military system cannot be achieved unless social scientists gain access to the reality of the system through appropriate techniques and methods of inquiry coupled with theoretical and conceptual frames of reference. Therefore, many problems are likely to remain below the surface unless critical incidents occur that enable social researchers to get access to the scene. An indication that some multicultural or multi-religious tensions do indeed exist in the Indian armed forces – despite all socialization and homogenization efforts – may be provided by the following recent newspaper report.

On 9 January 2006 the Army headquarters put on record its objection to the UPA government panel's questions on Muslims, and their roles in the armed forces. Sending the list of Muslim officers and their ranks, Army headquarters wrote to the Defense Ministry that the collection of such data could affect troop morale, as it could play into the hands of those who wanted to give it a particular political flavor. Army chief General J.J. Singh, who told a newspaper, 'It is not the Army's philosophy to discriminate or maintain such information', echoed this. He stressed that 'we are equal-opportunity employers. We strive to take people on certain standards after which only merit takes them forward. We do not bother about where they are from, their faith or language'.[17] Clearly, General Singh wants to downplay issues that indeed may exist in the organization, and – fully in line with the socialization practices referred to earlier – he emphasizes the unity aspect of the Indian armed forces. And he is right to do so, at least from his perspective. On the other hand, he may be paying too little attention to issues that are likely to claim his attention (or his successors' attention) in times to come. Overestimating the impact of a – possible – problem may be as precarious as underestimating its – possible – impact.

Soldiers are traditionally taught not to air their personal troubles too freely in public, and it must be said to their credit that this tradition is usually observed. However, in talking to an officer or his family one gets the impression that there are many perplexities for them these days. The main issues seem to be the early ages laid down for retirement, poor prospects of promotion to the higher ranks, separation from families because of service conditions, and difficulties in the education of children. Problems with respect to diversity issues in the armed forces are not mentioned very frequently, which clearly is an indication of the military's successful efforts to unite India's diversity. Nonetheless, military social science in India, if given the space that is normal in a maturely developed democracy, might raise several issues:

1 What distinctive methods can be formulated for the study of Indian military system in view of its secretive nature, non-accessibility and closed character?
2 What are the structural and cultural attributes of the military system that can be analyzed with regard to the diversity of its diverse workforce?
3 How has the process of modernization influenced the military system and how has the military system influenced modernization?
4 What are the dimensions of motivation, aspiration, value orientation, morale and commitment of the diverse categories of military personnel?
5 What is the relationship of civil society with the Indian military system?
6 What are the roles of the caste system and religion in the Indian military system?
7 What is the role of the media in the opinion-building process in India, as far as the armed forces are concerned?

A final word

The changing nature of civil–military relations in Asia has been a focus of attention for any serious scholar of peace and security. Military professionalism is an influential but controversial concept in the study of civil–military relations. The Indian armed forces have been unique among those in the developing world in several ways, since the historical background and political context are different in each country. Yet the framework enables us to study some of the trends of institutional development in Indian military organization. Since the relationship between the military and society in each social system also reflects national and cultural considerations, the need of the hour is to understand developmental trends over time through imperial research findings.

In contrast to what is found in the Western world, where we find the Cold

War bringing changes in the security concerns, leading to a formal redefinition of major missions, the Indian Army still focuses on what Moskos calls the 'defense of the homeland'. At the same time it participates in international peacekeeping operations and other humanitarian operations. But if we try to look into the military functions today, we definitely find some minor shift from primarily war fighting or war deterrence to military deployments for peace and humanitarian purposes. This may be due to a change in the perception of the threat from Pakistan and China. The shift is towards peace missions and humanitarian activities by the armed forces, and simultaneously we find some changes in public attitude and opinion.

To conclude, one can say that a system that is responsible for the maintenance of the territorial sovereignty of a country and that supports the internal security system obviously has paramount significance for sociological inquiry. Cross-national researches should be undertaken so that social, psychological and cultural aspects could be examined on a continuous basis. Sociologists should be able to assess the deeper aspects of resentment, frustration and aspirations in order to understand different societies in a broad framework. The Indian armed forces are possibly the third largest in the world and the Indian Army is the second biggest. Its long tradition, its set ways, its secular nature, its continuing ethos and strong sense of attachment to the regimental system give it a true corporate identity based on an approach to life that is centered around the military profession. The question is: how long will these values continue to hold, in a rapidly changing civil society? And – in the words of Amartya Sen again – how long will the India armed forces be able to foster their history, culture and identity by successfully combining their 'internal pluralism' and 'external receptivity'?

Notes

1 S.P. Cohen, 'Civilian Control of the Military in India', in C.E. Welch, Jr (ed.), *Civilian Control of the Military: Theory and Cases from Developing Countries*, Albany: State University of New York Press, 1976, pp. 43–64. See also R.L. Schiff, 'Concordance Theory: The Cases of India and Pakistan', in D. Mares (ed.), *Civil–Military Relations*, Boulder, CO: Westview Press, 1998.
2 A. Sen, *The Argumentative Indian: Writings on Indian History, Culture and Identity*, London: Penguin Books, 2005. Sen makes the telling point that since the Upanishadic or Mahabharata times, Indian thought has been characterized by arguments, disputations, questions and dialogues. One often tends to think of science and the scientific temper as Western, and brought to India by the colonial powers. Sen demolishes this thought, and points to India's long-standing tradition of reasoning and argumentation, leading to imperishable contributions to the various sciences.
3 Major-General (Retd) I. Cardozo, *The Indian Army: A Brief History*, New Delhi: Center for Armed Forces Historical Research, United Service Institution of India, pp. xii, xiii.
4 K. Kuldip Singh, *Indian Army*, New Delhi: Manas Publications, 1998, pp. 82–83.
5 I. Deva Shrirama, *Society and Culture in India*, New Delhi: Rawat Publications, 1999, p. 32.
6 S.P. Cohen, *The Indian Army: Its Contribution to the Development of a Nation*,

Berkeley: University of California Press, 1971, p. 49. See also T. Barkawi, *Globalization and War*, Lanham, MD: Rowman & Littlefield, 2006, p. 68.
7 It is among the few in the developing world with a continuous and unbroken history going back at least 200 years. It also policed the British Empire from the Middle East to China in the nineteenth and the first half of the twentieth centuries. It played a major part in World Wars I and II in Europe, Africa and Asia. See Barkawi, *Globalization and War*, especially ch. 3.
8 Muthian Alagappa (ed.), *Military Professionalism in Asia*, Honolulu, HI: East–West Center, 2001, p. 19.
9 Ibid., p. 20.
10 Lt Gen. A.M. Sethna, *Tradition of Regiment*, New Delhi: Lancer, 1993.
11 Several accounts exist of this period. Perhaps the best in this genre is the one by a distinguished Indian general and former vice-chief of the Army Staff, Lt Gen. S.L. Menezes: *Fidelity and Honour: The Indian Army from the Seventeenth to the Twenty-first Century*, New Delhi: Penguin Books India, 1993. See also Barkawi, *Globalization and War*.
12 Cohen, *The Indian Army*. See also Alagappa (ed.), *Military Professionalism in Asia*, p. 19.
13 Cardozo, *The Indian Army*, pp. xii, xiii.
14 S. Gupta, 'Kitne Musalman Hain?', *The Indian Express,* 11 February 2006.
15 Leena Parmar, *Society, Culture and Military System*, Jaipur and New Delhi: Rawat, 1994.
16 Col. D.S. Randhawa, 'Women Officers and Work Environment: Indian Perspective', *USI Journal*, July–September 2005.
17 S. Gupta, 'In Jan, Army HQ Said: Such data bad for morale", *The Indian Express*, 14 February.

9 Diversity in the Israel Defense Forces

Edna Lomsky-Feder and Eyal Ben-Ari

In this chapter we analyze the manner by which the Israel Defense Forces (IDF) 'manages' social and cultural 'diversity'. 'Diversity' as such is not a central issue, as it is in the armed forces of other countries, where the formulation of policies and arrangements for 'minorities' must be understood as part of a much wider rethinking about the future of the armed forces. Understanding the management of diversity in the IDF necessitates taking into account the ongoing debate about its continued role in the definition of Israeli 'nationhood' ('the people's army') and its evolution into a small, compact force based on professionals (a 'professional army'). It is within this wider debate, and its organizational and institutional derivatives, that the administration of diverse groups and demands should be seen.

Multiculturalism, diversity and politics

Against the background of the other contributions to this volume, we should state that the IDF does not use a rhetoric of 'managing diversity', but rather one of how new immigrants are 'assimilated' and how special populations are 'managed'. Moreover, in modern Israel the term 'minorities' relates to non-Jewish groups (Arabs and Druze, for example), while 'ethnicity' is used to refer to various Jewish groups.

While the terms 'diversity' and 'multiculturalism' are part of a public discussion about the politics of identity, they are not especially salient in regard to debates about the Army. The very manner by which the question of managing diversity is phrased, say the American example of managing women and blacks in the military, implicates a set of premises centering on a notion of integration as a normative ideal, and a practical possibility. What lies at base of this view is a particularly American liberal-democratic multiculturalism marked by respect for alternative ways of life and a belief in the basic equality of individuals. It is perhaps these kinds of assumptions that lie at the base of the US military's continued attempt to integrate various ethnic groups into its ranks.[1] But such a system may encounter difficulties in handling certain kinds of diversity. For example, how do the military act in a system based on such assumptions with groups that do not respect the basic equality of 'others' and want to impose their lifestyles on others (for instance, Christian fundamentalists in the United States or Muslim ones in France, who would enter the armed forces but be subject to the authority of their own religious leaders)?

The difficulties the Israeli case evinces differ even from this scenario. Encapsulated in the impossible combination of Israel as both Jewish and democratic, this society poses a completely different model of social and cultural diversity. First, while Israel is, like the United States, an immigrant society, it is (perhaps like Portugal or Japan) a uni-national state, based on Jewish nationalism. Because in this society 'Jewishness' is the definer of membership, minorities (by definition) can never achieve full membership. Second, Israel has a number of major groups that do not accept the democratic ground rules (for instance, the distinction between church and state) and would like to urge their way of life on all Jews. Third, these problems are amplified by the still strong ideological stress of the IDF as a 'military melting pot' and the practice of compulsory service, implying that the IDF must contend with the inclusion of such groups within its organization.

The 'holy quartet' of hegemony and its contestations

Forged as a force charged with securing Israel as homeland for Jews, from its beginnings the IDF was conceived of as 'the people's army', a veritable citizens' army. It was seen as the central symbol of the identity and solidarity of the new state and as a means to create the new, active Jew, radically different from the diaspora Jew.[2] But it was a rather clear picture of 'the people' that was encapsulated in this image. It was based on a 'holy quartet' of interrelated features: Jewishness, masculinity, military service and collective membership. To put it rather simply, but not incorrectly, according to this hegemonic conception, to be an Israeli in the 'full' sense of the word implied that one had to be Jewish, male, serving in the military and then granted full membership in the Israeli collectivity. It was on the basis of these criteria, in turn, that power, status and other resources were allocated to serving men. To be sure, there were always women and non-Jews in the IDF, but their position in the military was always secondary and subordinate to that of those men epitomizing the holy quartet of ideals. Indeed, the link between Jewishness, gender, military service and membership was so taken for granted in Israel until the 1970s that it can be called hegemonic.[3]

The first decades of the country were the heyday of Israel as a 'mobilized society': the state, primarily through the military, mobilized various social groups towards its own ends. By the same token, it sought to control social exclusion and inclusion implied by the 'holy quartet'.[4] For example, this model implied the exclusion of non-Jews from military service while demanding the acculturation and assimilation of Jewish immigrants to it. The role of the IDF as a 'military melting pot' (for Jews) was expressed via such activities as Hebrew language and Jewish history classes for immigrants.[5] Moreover, the assumption was that it was only in the army that all the strata of society could be subject to 'objective' criteria for advancement.[6] Citing a number of cases in which appeals were made to the military to enlist specific social groups so as not to deprive them through non-enlistment, Kimmerling contends that the draft was in effect a mechanism for fusing the collectivity's informal boundaries with the formal ones of citizenship.[7] As a result, a social group's demand for enlistment was a demand for full social membership.

Recruitment served both as a mechanism for building a hierarchy of social groups and as an indicator of this hierarchy.[8] Military service created a hierarchy based on access to conscription and on type of conscription. Where non-Jewish groups such as the Druze or Bedouins were allowed to serve, they were relegated to secondary status as citizens. Similarly, for most non-conscriptable groups – primarily Palestinian citizens of Israel – non-conscription was (and still is) used as an indicator of marginality or partial membership in the collectivity and as a mechanism for legitimating their partial exclusion.[9] What eventuated was a complex model of varying degrees of 'full' membership: at the center are serving Jews, with men in a superior position to women; more peripheral, but still related to the center, were Druze and Bedouins who served in the military; and at the outer edges of this structure were a range of groups who do not participate (Arab Israelis, handicapped people, some women, and ultra-religious Jews). High placement on one of the scales could offset a lower standing on another. To put this point by way of example, the high rank and combat experience of a Druze officer could offset his relative marginality in terms of national affiliation.

It is in such a light that the so-called full-fledged enlistment of women should be seen. We say 'so-called' because they were recruited in a manner that excluded them from the centers of power in the military. A number of scholars, such as Kimmerling and Izraeli, have explained how military service was (and still is) linked to distinct patterns of gender, power and inequality.[10] Given its basic male-orientedness, it is not surprising that the IDF has served to marginalize Jewish-Israeli women from the most important Israeli institution. The consequence of this situation has been the reinforcement of women's general marginality in Israeli society, and their exclusion from the most important societal discourse, that of 'national security'. The hegemonic model thus implied the establishment and reproduction of the lower status of such groups as women and 'minorities' such as the Druze and Beduin.

Yet this hegemonic ethos and its concrete ramifications began to erode after a few decades of the state's existence. It is especially since the beginning of the 1970s that Israel has seen the emergence of new voices contesting the 'holy quartet'. The gradual erosion of this model, however, has not spelled the total disappearance of the values and sentiments associated with it. What seems to be happening is that competing worldviews and assertions of Israeli identity and peoplehood are finding greater public expression. From our perspective, this debate has led to the emergence of a number of 'contested arenas' in and around the military and military service. We chart out the social locations from which the hegemonic model is being contested, because it is in reaction to these critiques and demands that the IDF has developed various policies and practices to deal with 'diversity'. Let us provide a few examples.

One viewpoint is that of the more radical part of the national-religious camp who, instead of providing a civic definition to military service, give it a messianic-religious one.[11] In providing such a definition, this group questions the very basis of military service as a duty of citizens and represents the threat of recruiting soldiers with fundamentalist orientations into the military. These issues have come to a head in and around the controversy over the role of such soldiers in the forced evacuation of settlements in the occupied territories. This controversy arose when a number of rabbis stated that their authority – and their dictation not to give up any area of the Land of Israel – stood higher than that of military commanders.

Another view, rooted in a liberal version of feminism, contests the masculine side of the 'holy quartet'. Concretely, this criticism is carried by many of the feminist movements in Israel and centers on the link between masculinity and military service.[12] In essence, the demands of this group have been to open up what are exclusively male military occupations to women. Here, the stress is less on a questioning of the hegemonic model than on an insistence on its extension to women.

Yet another outlook is the criticism that for non-Jews, military service does not in effect assure full social membership and citizens' rights. This view is held by Druze (whose males must serve, according to law) and Bedouin (who volunteer for the IDF). Their critique centers on how, despite serving in the military, they are granted only second-grade membership. At times explicitly and at other times implicitly, many of their appraisals underscore how the IDF is basically a Jewish army and how non-Jews can only be granted secondary status within it. They, like women, are trying to extend the boundaries of access to rewards accruing service and do not undermine the link between military service and citizenship.

Big army – small army: the central debate about the IDF

In contrast to most of the armed forces of other industrial democracies, the IDF is still based on conscription and thus has much less of a problem in competing with civilian workplaces for recruitment and retention of personnel. Partly as a result of changes in its threat environment, partly because it faces the same sort of social problems as the armed forces of other societies, and partly as a result of transformations in Israeli society, during the past two decades the IDF has been grappling with a set of major challenges that center on its very character. A host of commanders, commentators and other scholars have been arguing the IDF, long a paragon of 'the people's army', should develop into a small professional force modeled on the armed forces of other industrialized democracies.[13] The argument is that in order to meet Israel's external threats, external demands for economic rationalization, the ability to retain highly qualified people and the ability to respond to changing attitudes to military service, the IDF must be transformed into a force that is technologically advanced, compact and flexible.

Concretely, such calls have resulted in what can be called a 'silent revolution' in the organization of the IDF. For example, the IDF has decreased its role in education or programs for disadvantaged youths. The IDF now asks for remuneration for some of the non-military tasks it carries out (such as sending auxiliary teachers or medics to civilian ministries). But probably the most important change has occurred in what, in effect, is a downsizing of the two main components of the IDF: the conscripts and reserves troops that bear the brunt of active service. Today, only about 55 percent of a given cohort of Israeli youths are recruited for compulsory service, and among reserve soldiers not only has the age at which they are released from duty been lowered, but fewer and fewer reservists are mobilized each year.[14] In addition, more regular units have been established to take the place of the reserve units and increasingly deployed in former reserve-allocated deployment areas in the West Bank and on various of Israel's borders.[15] Finally, the IDF has continually needed to show how it is rationalizing in its struggle for budgets. Indeed, within the IDF some of the hottest buzzwords include such terms as 'privatization', 'outsourcing', or 'excellence in management'.[16]

Yet the move to a small, professional army is not uncontested. The transformation of the IDF into such a force goes against the grain of the (slow-changing) mainstream assumption that the obligation of national service is related to the very survival of the Israeli nation-state and a means through which to gain membership in 'the collectivity'. Demands for transformation into a professional force have met with resistance. Arrayed against people pushing for the evolution of the IDF into a smaller army is a formidable array of groups and individuals (many within the military) who stress that the IDF's nation-building roles have not ended. These people, out of commitment to a citizen's army, out of a wish to maintain the legitimacy of the IDF or out of a continued allegiance to the ideals of Israel's founders, maintain that Israel's military must continue the practice of compulsory service, to integrate various groups and to provide a basis for citizenship. In addition, as commentators have noted, if compulsory service is ended, then the top 30 percent of the recruits – those most needed for the technologically advanced military – will not volunteer for the IDF.

This debate about the contemporary IDF is thus marked by a deep-seated tension between the twin representations of 'the people's army' and a professional force. It is rent with conflicts and ambivalence about the IDF's future development. It is into the broad contours of this debate that discussions about the place of Israel's diverse groups enter. As we show, because these groups offer distinct critiques of the definition of 'Israeliness', they present the IDF with distinct challenges to the ways it mediates the tension between 'the people's army' and a 'professional army'.

Different uniforms for the people

We now turn to examine a number of specific groups that the IDF itself defines as being culturally diverse and thus necessitating special treatment. In this respect our analysis is based on a distinction between four analytical dimensions.

First is the distinction between marked and unmarked categories. Thus, for example, while outside the IDF the ethnic issue – often dubbed the ethnic cleavage – between Mizrachim and Ashkenazim (Jews from Middle Eastern and from European origins respectively) is perceived to be central to the creation of social hierarchies, within the Army it is not alluded to formally. Moreover, in contrast to the civilian sphere, where this issue prompts many interpretations and debates, within the Army there is a very strong opposition to talking about this distinction in terms of diversity. Within the Army the governing conception is thus of an 'ideal' Israeli surrounded by 'special' groups that are explicitly and consensually defined as different, because of their 'newness' to society (immigrants), ethnic visibility (such as Ethiopians), possession of a distinct lifestyle (religious Jews) or distinct social positions (Arabs and women). The second dimension is the kind of rhetoric surrounding the administration of a specific group: the specific 'framing' of the group's character or requirements. The third dimension concerns the actual organizational practices put into place in order to 'manage' members of a specific group in the IDF. The fourth and final, if often overlooked, dimension is that of the informal arrangements that have evolved in regard to the group.

At the outset let us mention two points related to the potential of the IDF for managing diversity. First, the IDF is not monolithic. Israel's forces are internally complex, with different groups, frameworks and individuals holding diverging views about, and capabilities for, changing the status and positions of various groups. Second, despite changes in recent years, Israel's Army is still rich in resources: it is a powerful organization. These traits mean that the IDF can (and does) react to different situations in multifarious (sometimes contradictory) ways, to engage in trial and error, and to act both in its long-term and its short-term interests. Given that the IDF is a fighting army, however, experimentation carries high risks. Thus, it is not surprising that women – 'dispensable' as they are (from an organizational point of view) – are the group in connection with which the greatest deal of experimentation has taken place.

Analyzing the rhetoric, policies and arrangements found in the IDF, we found four main practices for dealing with diversity:

- support – the existence of provisions, frameworks and special roles that provide for members of the group in social, cultural, psychological and instrumental ways;
- exclusion – of the group's members from prestigious roles and blocking their prospects for promotion;
- controlled assignation – control and supervision over the entry to 'sensitive' roles, allowance of the development of cultural enclaves and, often, separate judicial treatment;
- screening information – control of the manner, degree and timing of release of information dealing with a group.

The combination of these practices leads to different patterns of managing diversity. Yet the actual pattern is an outcome of the tension mentioned before between a 'people's army' and a 'professional force': on the one hand, the main considerations in making policy are based on the personnel needs of the IDF while on the other hand, the treatment of different groups is derived from the 'threat' that it poses for the place of the IDF as a people's army.

Immigrants: supportive integration

In Israel, Jewish immigrants are automatically granted citizenship and are liable for compulsory military service. Long an explicit category towards which policies have been formulated, handling of immigrants has been based on an assumption of the IDF as a 'military melting pot'. The immigrant groups that have arrived since the end of the 1980s (800,000 from the former Soviet Union and about 50,000 from Ethiopia) pose two kinds of problems. First, as with the immigrant waves of the 1950s, their sheer number poses difficulties in processing large aggregates of individuals, many of whom do not fit the needs of a professional army. Second, immigrants from the former Soviet Union often threaten the military's dominance as an arena for actualizing membership.[17] The discourses centering on immigrants revolve either around their integration or around their unique human capital for improving the functioning of the IDF.

The formal system serving the needs of new immigrants includes classes in the Hebrew language, Israeli history and Jewish tradition; pre-Army courses; financial aid, including supplementary salaries, welfare benefits or housing allowances; specialized departments charged with dealing with 'Russians' and 'Ethiopians'; designated personnel (invariably women) at unit level dealing with immigrants; and courses and booklets devoted to raising the 'cultural sensitivity' of commanders.[18] It should be noted that many policies are based on an assumption that the status of immigrants is transitory and that 'all' the IDF need do is ease the transition of immigrants to full Israeli status. Yet the most important if unstated policy regarding immigrants is one dictating that the IDF focus its investments only on 'worthwhile' groups in terms of the needs of the professional army. Thus, many immigrants are granted lengthy deferments and some exemption from military duty altogether.[19] The high level of many such recruits has allowed the IDF to raise the level of some roles like drivers or mechanics. Also, so as to sidestep thorny issues related to the 'Jewishness' of the immigrants, the military has instituted special burial procedures. Finally, the IDF has been central in allowing thousands of immigrants to undergo formal conversion to Judaism.

Probably the most important informal arrangement that has evolved (and it is a radical departure from earlier periods) is the unofficial 'permission' immigrants have been granted to maintain a 'local' culture within units: to speak Russian between themselves but strictly use Hebrew in all operational and organizational matters;[20] to read their own Russian language books and newspapers; and to be called (in a good-natured manner) such things as 'the Caucasians' or 'Russian mafia'. Such unofficial permission again is actually seen as temporary, as these groups are perceived to be in a transitory state on the way to full-fledged membership in Israeli society. In this respect, however, while Ethiopians are dispersed throughout the IDF and do not develop their own culture, it is among immigrants from the former Soviet Union that such informal cultures have developed. Yet for many of these people, military service does not automatically lead to the adoption of wider Israeli identities. For many ex-Soviets, entrance into Israeli society entailed by service does not necessarily include sentiments of commitment or identification. Rather, Army duties are seen to be part of a 'transactional game' in which individuals weigh the opportunities involved in terms of their self-interest.

It appears that over the past decade the IDF has been rather successful in integrating immigrants from the ex-Soviet Union, who are now all graduates of the Israeli educational system. This success is evident in their high recruitment rates and the fact that more and more of them are volunteering for combat units and officers' courses. On the other hand, the prevailing sentiment is that the case of Ethiopians has not been as successful, and this is often attributed to their being culturally different. During the past half-decade the IDF has been placing special emphasis on this latter group as part of its embodying the notion of a 'people's army'. This move has expressed itself in affirmative action plans for promotion and special programs to accompany individuals even before conscription.

To conclude, there is a well-developed system for preparation and direction of immigrants based on the needs of the IDF, which also permits the development of local cultures. When the Army drafts the majority of immigrants but invests only in a minority that serve a full three years, it gains external legitimacy and ensures that its requirements are met. In a word, this group is included with minimal systemic 'noise' and maximum profit (internally for professionalization and externally for legitimation). They are, in sum, mobilized in the name of 'the people's army' for the needs of the professional army.

Women: forced integration

While women have been an explicit category recognized by the IDF, in reality they have been overwhelmingly posted to various clerical, support or 'caretaking' (teaching or social welfare) roles. With the expansion of the IDF in the 1970s women were allowed to take up some positions that had previously been the exclusive preserve of men (for example, tank or artillery instructors). The reasoning at base of this change was the idea of freeing more men to participate in combat roles. Yet this change was also the result of external pressure placed on the IDF by the political force of the women's lobby in the Knesset. Today, the thrust of the pressure applied on the IDF in this respect centers on opening up especially combat roles to women and assuring equal promotion in the force. Women seem to question the basically male-oriented nature of the Army and its consequent biases.[21]

Two kinds of rhetorics have been used to justify the treatment accorded women and restrict their access to important roles. While the first centers on their bodies as weaker than men's, the second involves circular reasoning focusing on the 'profitability' of investing in them.[22] Thus, even when some senior commanders contend that women can undertake the same roles as men, they contend that, given their short period of service, they are an unprofitable investment because prospects for long-term payback are so weak. In practice these attitudes express themselves in the assignation of the majority of women to support positions.

In addition to the shorter period of service for women (twenty-four months as opposed to thirty-six months for men), policies directed towards women also include rather easy exemptions, practically no reserve duty, and, for those women whom the Army considers to be of value, a system of pre-service preparation (effectively lengthening their service). Finally, in the past few years the IDF has recognized and come to act upon sexual harassment by instituting new regulations and schemes to raise the awareness of commanders to this issue. Accompanying these visible measures are a host of unstated (if very real) biases in the application of ever stricter criteria restricting the number of women in the IDF (albeit as part of the downsizing of the military); no real encouragement for women to pursue long-term military careers; and glass ceilings in promotion (most severe around the rank of major and lieutenant-colonel).[23] On a more informal level we find evidence that the IDF's male-oriented culture results in a marginalization of women in the day-to-day life of military units.[24]

The integration of women is still limited, and the most significant struggle is waged over incorporation into the last bastions of male exclusivity, combat roles. As a consequence of external legal and political pressure and the internal pressure of some commanders, the IDF has been forced to integrate women into such roles as pilots' or naval commanders' courses, the artillery and light infantry. But opening up selected roles has not meant sweeping admission, because the military fought a rearguard action on a point by point basis. Changes in policy have emerged when the IDF surrenders to external pressure in the name of the ethos of 'a people's army'. In reality, however, the Army has harnessed women to the needs of a small professional army only in a relatively peripheral manner (in contrast to the American forces). If, however, we appreciate that the IDF is not internally consistent in its reactions, we can understand how women have used it as an agent of change. In the case of opening up job categories to women, a coalition between external women's movements and groups within the IDF (mainly advisers on women's affairs and some senior officers) has led to some success.

At the same time, however, the past few years have seen a number of interesting developments. First, the position of Chief Women's Officer has been abolished and instead a Consultant to the Chief of Staff for Women's Affairs has been established. The holder of this role is charged with encouraging the promotion of women in the organization. Second, some units have realized that actively recruiting women into their midst allows them to raise the general level of their troops (a prime example is technicians' roles). Third, at the level of public rhetoric the IDF stresses the equality of women and men, but in concrete terms this means women adapting to the requirements of the organization rather than the military adapting to the needs of women. Finally, the discourse on women as objects to be protected has been transformed by some religious groups into a discussion about limiting the activities of women in the Army so that religiously observant men can serve there.

In sum, while women are recruited on the basis of assumptions derived from 'the people's army', they are considered to be dispensable and are only partly harnessed to the needs of the small professional army. Organizational practices continue to reproduce their marginality alongside a marketing effort that plays up their entry into new roles as a major shift in policy. The continuity of exclusion is hidden beneath a veneer of change.

'National minorities': partial integration

Long an explicit category recognized by the IDF, 'national minorities' comprise Druze and Circassians who are conscripted, and Bedouins and (a small number of) Arab Christians who volunteer.[25] While not always clearly expressed, talk about minorities revolves around two rhetorics. The first is an organizational 'orientalism' in which Jewish troops endow these minorities with special characteristics 'suitable' for special military roles. Bedouin, for example, are often reasoned about by linking their existence 'closer to the land' to their more developed senses of sight, hearing and smell and therefore to their suitability to be good trackers. The Druze are often remarked about in terms of being 'natural' warriors suited to the infantry. The other rhetoric, often expressed during ritual occasions as funerals, is that of their being 'brothers in arms' to Jewish warriors (a 'brotherhood' which, ironically, excludes women) or partners in a 'covenant of blood'.

The formal arrangements instituted in regard to minorities include assignation to special units and administration by specialized personnel departments.[26] These arrangements are complemented by the ethnicization of their military role. In tangible geographical and abstract symbolic terms they deal with the 'borders' of Israel. Thus, for example, as trackers they are an integral part of patrols along the country's international boundaries. Similarly, the Druze are central core of the system by which policing of Palestinians takes place (in the military administration of the West Bank, in security prisons and in the border police).[27] In addition, many members of the special minority units are 'subcontractors' for other units: while trackers belong to the Bedouin battalion, they are often sent in small groups to aid other regular battalions. Finally, within the general Jewish religious framework of the IDF, they are allowed to carry out their specific religious arrangements. An unstated but formal procedure, which is still in effect, is the exclusion of minorities from certain 'sensitive' units like the Intelligence Corps or (until the early 1990s) the Air Force. This policy is related to what Peled calls the 'Trojan Horse' dilemma: '[C]an members of ethnic groups whose loyalty to the state is questioned by the government and the general public be recruited, trained, integrated, and promoted within the armed forces?'[28] In practice, this dilemma expresses itself in the principle of 'divide and rule'. Take the formal distinction between the Druze, Circassians, Bedouin and Christian Arabs, which ensures that there is no development of a 'pan-Arab' identity based on their common Arab culture. Indeed, these distinctions are based on distance from Arab Muslims, who are granted the least trust. This division is further strengthened by the differentiation between compulsory duty for Druze and Circassians, as opposed to a voluntary one for Bedouins. Furthermore, while they are granted permission to use Arabic among themselves, Hebrew must be used in all operational and organizational matters.

Israel's 'national minorities' have become increasingly vociferous since the 1970s in pressuring for access to all units and promotion, and in petitioning for full citizens' rights by virtue of military service. The IDF's reaction is based on how these groups meet its professional needs and involves a mix of phased integration, playing up individual success stories, and occasional tokenism. After the 1973 war, when minority troops 'proved' themselves on the battlefield and the IDF needed infantrymen, barriers were gradually removed.[29] Yet it is our impression that these groups suffer lower promotion rates, and many roles are still off-limits for them. Limited progress has been made, though, as the first Druze to graduate from the naval commanders' course took a command post.[30]

Externally, their demands have been for greater political participation and allocation of state resources.[31] 'We are not ethnic soldiers,' they seem to be saying, 'but full citizens of Israel (according to the hegemonic model).' Additionally, the IDF conceals cases of men who do not want to be mobilized. While occasional reports of Druze shirking military service are published, there are no real data about the extent of this phenomenon.[32] Indeed, the tensions within these groups about serving in the IDF are reaching new heights. The main tension seems to be between an attraction to military service for the economic benefits it represents and trends towards a strengthening of wider Arab and Palestinian identities among these groups, and the feeling that they are second-class citizens despite serving. As in the case of women, change has been brought about through a coalition between external group leaders and commanders in the IDF.

Being outside 'the (Jewish) people', they are mobilized for the needs of the professional army. They are directed into important occupational niches for instrumental reasons under strict supervision. Their continued recruitment, then, fits the move to a professional army: they are specialized soldiers, have high motivation, and tend to stay on in the permanent force as professionals. But, given the limits on their integration, it is no surprise that Peled decided to call his chapter on minorities in Israel 'Between inclusion and exclusion'.[33]

National-religious Jews: controlled integration

While the religiously observant troops comprise no more than 10 to 15 percent of the IDF, their significance is wider, since they are important for outlining the boundaries of the military and obtaining support for its actions. In this subsection we deal with the national-religious group (the orthodox Zionists) and not with the ultra-orthodox Jews.[34] The participation of Jewish fundamentalists in the IDF involves acute questions: where does the allegiance of such soldiers lie (with rabbinical authorities or military commanders?) and will such soldiers enforce Jewish orthodox customs on their fellows? The wider threat they pose lies in replacing the definition of military service as a citizen's duty by a religious one. The background to these apprehensions is the different values of the religious and secular groups in Israel. The formal rhetoric plays up their great contribution by virtue of being high-quality and very motivated troops. In informal rhetoric they represent the possibility of a fundamentalist takeover of the IDF.

The formal and informal arrangements for handling this group emanate from their special needs and threats they pose. Alongside regular recruitment, many of them are mobilized into special units, where they alternate between regular service and religious study.[35] This arrangement, allowing the IDF control, is accompanied by strong socialization to military values (as with all recruits), and selection only of those persons who fit the structure of authority of the IDF. Finally, while there are internal studies of this group, the IDF releases almost no information regarding their numbers in elite units or different ranks among officers. The feeling of threat, however, is mutual. Military service is seen by many religious people as an experience drawing away youngsters from religion. Thus, serving in homogeneous units allows members of this group to create their own cultural enclaves.

In addition, conscripts from orthodox homes minimize the threat of military service through the propagation of handbooks for religious conduct, studying in pre-conscription religious colleges or allowing religious women to serve outside the IDF in alternative avenues for national service.[36] Stuart Cohen, however, maintains that the threats of the national-religious sector are exaggerated: for example, the national-religious camp is itself rent by deep divisions and as yet there have been relatively few refusals of military authority in the name of a higher rabbinical authority.[37] Thus, out of the tens of thousands of troops who participated in the disengagement from the Gaza Strip in 2005, only sixty-three soldiers refused to carry out their assignments. But what is important in this respect is that while the IDF was 'victorious' in the disengagement, and most religious soldiers did not go against the state, debates within and outside the Army about the special routes and arrangements for these soldiers and the advantage of concentrating them in special units have now emerged.[38]

In addition, Cohen has shown that the long-term prospect of the 'takeover' of the IDF by religious commanders is not realistic, because their communities are themselves 'greedy institutions', like the military.[39] For instance, a young religious officer contemplating a military career faces a difficult choice because of strong demands for him to return to religious establishments of higher learning. In addition, despite popular and scholarly images to the contrary, this camp is heterogeneous in the kinds and degrees of commitment to military service found among its young people. Thus, Cohen is careful to talk about patterns (in the plural) of service among such youths.

In sum, since they are seen as very suited to the needs of the professional army, they are mobilized – and very actively mobilize themselves – in the name of 'the people's army'. But because they represent threats to the character of the IDF, they are controlled through mechanisms of socialization, dispersal or insulation.

Conclusion: profiles and models

What are the wider lessons of the Israeli case? We end with a number of suggestions about how this case enables us to build a wider model of policies and practices used in regard to diverse groups.

First, management of information is central for handling relations with the environment because it allows the Army some control over images of how it manages different groups. Thus, for example, the IDF is rather hesitant about releasing information about such subjects as the number of ultra-orthodox Jews exempt from service, the promotion of Druze to senior positions, or the dispersion of youths from the national-religious camp among units and roles. Conversely, Israel's armed forces have become increasingly sophisticated in using the media to play up favorable stories related to women and national minorities.

Second, an unintended consequence of the integration of minority groups has been the creation of internal pressure groups: special frameworks for dealing with minorities have become internal pressure factions for the groups they were set up to serve. Examples are the Bedouin Battalion or the personnel departments charged with immigrants who now pressure the Army regarding their groups' interests.

Third, the Israeli model is built on combinations of inclusion and exclusion, control and permission, support and disregard, and dispersal and isolation. Analytically, it allows us to isolate the major parameters that govern these principles and arrangements. Thus, specific profiles of managing diversity are related to the overall social and cultural 'logic' by which they are constructed. The external rhetoric used in regard to each group derives primarily from the notion of 'the people's army' and is geared to continued external support for the IDF, but the specific IDF treatment is an outcome of the 'internal' and 'external' threats that the group poses.

Finally, it is not possible to talk about the management of diversity without examining wider concepts of 'citizenship', 'integration', or 'pluralism' that form the cultural environment of a given military organization. Even if the IDF does not turn into a professional volunteer force, it will downsize and we shall see more and more negotiations between the military and social groups about the length and character of service.

Notes

1 G.H. Lawrence and Th.D. Kanel, 'Military Service and Racial Attitudes of White Veterans', *Armed Forces and Society*, 22, 2, 1995/96, 235–256.
2 O. Almog, *The Sabra: The Creation of the New Jew*, Berkeley: University of California Press, 2000.

3 U. Ben-Eliezer, *The Emergence of Israeli Militarism, 1936–1956*, Tel Aviv: Dvir, 1995 (in Hebrew); U. Ben-Eliezer, 'A Nation-in-Arms: State, Nation, and Militarism in Israel's First Years', *Comparative Studies in Society and History*, 37, 2, 1995, 264–285; S. Helman, 'Militarism and the Construction of Community', *Journal of Political and Military Sociology*, 25, 1997, 305–332; E. Lomsky-Feder, *As if There Was No War: The Life Stories of Israeli Men*, Jerusalem: Magnes, 1998 (in Hebrew).
4 D. Izraeli, 'Gendering Military Service in the Israeli Defense Forces', *Israel Social Science Research*, 12, 1, 1997, 129–166.
5 M. Lissak, 'Paradoxes of Israeli Civil–Military Relations: An Introduction', in M. Lissak (ed.), *Israeli Society and Its Defense Establishment*, London: Frank Cass, 1984, pp. 1–11; A. Peled, *A Question of Loyalty: Military Manpower Policy in Multi-ethnic States*, Ithaca, NY: Cornell University Press, 1998.
6 Y. Levy, 'Social Convertibility and Militarism: Evaluations of the Development of Military–Society Relations in Israel in the Early 2000s', *Journal of Political and Military Sociology*, 31, 2003, 71–96.
7 B. Kimmerling, 'Determination of Boundaries and Frameworks of Conscription: Two Dimensions of Military Relations', *Studies in Comparative International Developments*, 14, 1979, 22–41.
8 D. Horowitz and M. Lissak, *Trouble in Utopia: The Overburdened Polity of Israel*, Albany, NY: SUNY Press, 1989.
9 Z. Rosenhek, 'The Exclusionary Logic of the Welfare State: Palestinian Citizens in the Israeli Welfare State', *International Sociology*, 14, 1999, 195–215.
10 B. Kimmerling, 'Patterns of Militarism in Israel', *European Journal of Sociology*, 34, 1993, 196–223; Izraeli, 'Gendering Military Service'.
11 M. Feige, *One Space, Two Places*, Jerusalem: Magnes, 2000 (in Hebrew).
12 I. Jerbi, *The Double Price: The Status of Women in Israeli Society and Women's Service in the Military*, Tel Aviv: Ramot, 1997 (in Hebrew); O. Sasson-Levy, 'Military, Masculinity and Citizenship: Tensions and Contradictions in the Experience of Blue-Collar Soldiers', *Identities: Global Studies in Culture and Power*, 10, 3, 2003, 319–345.
13 S. Cohen, 'The Peace Process and Its Impact on the Development of a Slimmer and Smarter'Israel Defence Force', *Israeli Affairs*, 1, 4, 1995, 1–21.
14 *Haaretz*, 5 December 2003, 7 January 2004.
15 *Jane's Defence Weekly*, 17 November 2004.
16 *Maariv*, 14 August 1998.
17 E. Lomsky-Feder and T. Rapoport, 'Juggling Models of Masculinity: Russian-Jewish Immigrants in the Israeli Army', *Sociological Inquiry*, 73, 1, 2003, 114–137.
18 D. Novack, 'The Cultural Branch of the IDF Education Corps', in D. Ashkenazy (ed.), *The Military in the Service of Society and Democracy: The Challenge of the Dual-Role Military*, Westport, CT: Greenwood, 1994, pp. 76–79.
19 Cohen, 'The Peace Process', p. 4.
20 *Maariv*, 13 February 1998.
21 E. Levi-Schreiber and E. Ben-Ari, 'Body-Building, Character-Building and Nation-Building: Gender and Military Service in Israel', *Studies in Contemporary Judaism*, 16, 2001, 171–190.
22 Izraeli, 'Gendering Military Service'.
23 Ibid.
24 O. Sasson-Levy, 'Gender Performance in a Changing Military: Women Soldiers in Masculine'Roles', *Israel Studies Forum: An Interdisciplinary Journal*, 17, 1, 2001, 7–23.
25 Peled, *A Question of Loyalty*, pp. 166–167.
26 Ibid., p. 163.
27 H. Frisch, 'The Druze Minority in the Israeli Military: Traditionalizing an Ethnic Policing Role', *Armed Forces and Society*, 20, 1, 1993, 51–68.

28 Peled, *A Question of Loyalty*, pp. 1–2.
29 Ibid.
30 *Maariv*, 21 August 1998.
31 Frisch, 'The Druze Minority'.
32 *Haaretz*, 29 January 1998.
33 Peled, *A Question of Loyalty*, ch. 4.
34 N. Stadler and E. Ben-Ari, 'Other-Worldly Soldiers? Ultra-Orthodox Views of Military Service in Contemporary Israel', *Israel Affairs*, 9, 4, 2003, 17–48.
35 S. Cohen, *The Scroll or the Sword? Dilemmas of Religion and Military Service in Israel*, Amsterdam: Harwood Academic, 1997.
36 Ibid., pp. 61–63.
37 Ibid.
38 *Haaretz*, 14 November, 2005.
39 Cohen, *The Scroll or the Sword?*

10 Ethnic diversity in the British armed forces[1]

Christopher Dandeker and David Mason

In our contribution to the earlier volume *Managing Diversity in the Armed Forces*, we drew attention to an evolving policy agenda focused on the need to make the armed forces more representative of the society they served. This agenda had its clearest expression in the then recently published *Strategic Defence Review* White Paper (SDR)[2] and entailed a particular emphasis on the apparent under-representation of Britain's citizens of ethnic minority descent when compared with their presence in the population as a whole. In this revised discussion we first revisit our analysis of the arguments that underpinned the SDR commitments. Against this background we consider what progress has been made in meeting the targets set, and identify outstanding issues and emergent problems.

Representation, fairness and effectiveness

There are, characteristically, two kinds of arguments that are adduced in support of equal opportunities measures. They appeal, respectively, to considerations of equity and fairness, and to self-interest. With respect to the first, a lack of equity is typically seen to raise issues of social justice and citizenship that have implications well beyond the boundaries of particular professions or occupations. Partly as a result, since 2001 the Race Relations Amendment Act has placed a duty on all public bodies to take positive steps not only to eliminate discrimination but also to promote racial equality. In the case of the second, often characterized as the 'business case' for equal opportunities, the issue is placed firmly within the realm of self-interest rather than relying on altruism or equity, characteristically justifying action in terms of what the organization is thought to gain from taking action.

Both these kinds of arguments resonate with concerns made explicit in SDR. At the level of fairness and social justice, for example, the following point is made:

> We are ... committed to making real progress on improving our record on equal opportunities through tackling the complex web of underlying factors which have inhibited people from various backgrounds choosing to join us

in the past. We must ensure that those who join us make progress according to their talents and legitimate aspirations.[3]

From the perspective of self-interest, or 'the business case', representation is recognized to be one element in both sustaining public support[4] and improving levels of recruitment and retention. These arguments have meant that the armed forces are increasingly being led to re-examine a range of traditional practices and assumptions, an impetus reinforced by legal pressures flowing both from the UK legislature and from such transnational bodies as the EU Commission and the European Court. Such rulings have led to rapid and far-reaching changes in the position of women within the UK armed services such that 73 per cent of posts in the Naval Service, 70 per cent of posts in the Army and 96 per cent of posts in the RAF are now open to them.[5]

By contrast with this picture of rapid change, progress in increasing the recruitment and retention of members of ethnic minority groups has been slower. A series of embarrassing allegations of discrimination and harassment, together with an investigation by the Commission for Racial Equality,[6] combined with recruitment pressures to produce a greater recognition, on the part of the armed services, that the under-representation of ethnic minority groups was a problem.

Representation and citizenship

In its discussion of equal opportunities for members of ethnic minority groups SDR placed emphasis on the need to increase ethnic minority recruitment incrementally so that, eventually, 'the composition of our Armed Forces reflects that of the population as a whole'.[7] Thus:

> We are determined that the Armed Forces should better reflect the ethnic composition of the British population. Currently some 6 per cent of the general population are from ethnic minority backgrounds, but they make up just 1 per cent of the Services. This must not continue. We have set a goal of attracting 2 per cent of new recruits this year from ethnic minority communities for each Service. We want that goal to increase by 1 per cent each year so that, eventually, the composition of our Armed Forces reflects that of the population as a whole.[8]

The implication is clear: proportional representation is a worthy goal in its own right and issues of fairness and citizenship are at stake.

The business case: public image, recruitment and retention

The first manifestation of a business case reflects a broader discussion that has taken place in recent years on the need for the all-volunteer forces to remain in touch with the wider society on which they depend, despite trends that have weakened these civil–military bonds. These include the following:

1   The armed services are smaller and have a lower social profile in society as a result of downsizing and base closures.
2   Since the abolition of conscription in 1963 there has been a profound decline in the number of people who possess a direct understanding of military affairs. Such people include influential opinion leaders, who can play a significant role in the defending the defence budget and wider interests of the armed services, and 'gatekeepers' (those who can influence young people in their choice of careers).
3   The purpose of the armed forces is less clear than it was in the era of the Cold War, when a direct and palpable military threat to the United Kingdom was widely recognized. Consequently, it is rather easier than before for non-military demands on public expenditure, such as health and social welfare, to push the armed services further down the order of spending priorities.
4   The armed services have a unique culture because of the requirement to prepare for and conduct war and other military operations. Maintaining the legitimacy of the services' 'need to be different' has become more problematic in recent years because of social and legal changes. In a climate that is less deferential, the burden of proof can no longer be placed upon the proponents of change.[9] The onus is now on the military to prove that conforming to the changing norms and values of wider society (for example, with regard to the employment of women in all military occupations) would be likely to damage operational efficiency.

A key aspect of the significance of the armed services' public image is the way in which it impacts directly on recruitment and retention. From a demand-side perspective, the SDR identifies both improved recruitment and, in some ways more significant, enhanced retention as key parts of a solution to current problems of understaffing and overstretch in the services.

Representativeness and the business case

Commentators have identified a number of reasons why the objective of making the armed services representative of the populations they serve is a desirable one from a business perspective.[10] First, the pursuit of greater representativeness would improve access to a wider recruitment pool as the armed services compete with civilian companies for labour scarce in terms of both quantity and quality. In this context we can point to the fact that while ethnic minority groups comprise 7.9 per cent of the population of the United Kingdom and 9.0 per cent of the population of England,[11] most have a significantly younger age structure than the white population. Thus, the 'Mixed' group had, in the 2001 Census, the youngest age structure, with 50 per cent under the age of 16. Thirty-eight per cent of both the 'Bangladeshi' and 'Other Black' groups were aged under 16, and 35 per cent of 'Pakistanis' also fell into this age group. This was almost double the proportion of the 'White British' group, for which one in five (20 per cent) were under the age of 16.[12] As a result, ethnic minority groups constitute a significantly larger proportion of the age group relevant to military recruitment than they do of the population as a whole; and they will continue to do so for the foreseeable future. In addition, there is evidence that members of some ethnic minority populations are increasingly more likely than their white peers to remain in education after the age of 16, thus potentially providing the services with a useful additional pool of skilled labour.[13]

Second, it is argued that the armed services would benefit from the diversity of skills and backgrounds that a broader-based entry would produce. With the need for more intelligent and flexible service personnel likely to increase rather than decrease, such diversity is likely to prove an advantage in future years.

Third, the services could benefit from being seen to live up to the ideal of being an equal opportunities employer. While this may enhance their standing in ethnic minority communities, it is just as important to sustain the legitimacy of the armed services and thus the fount of good will among the general public. In any case, the legal pressure to conform to this ideal is real enough.

Progress towards the target of representativeness?

The first few years of policy yielded disappointing results. However, more recent data appear to show that the policy first enunciated in the SDR is beginning to pay dividends, with clear evidence of greater success in recruiting ethnic minority personnel in the most recent years for which data are available. On the face of it, the figures are beginning to approach, if not exceed, the government's targets and to approach the proportion of the population recorded by the Census as being of ethnic minority origin.

Table 10.1 contains data relating to recruitment and shows a broadly upward

Table 10.1 Intake to untrained strength of UK Regular forces by ethnic origin and service (%)

	1998/99	*1999/00*	*2000/01*	*2001/02*	*2002/03*	*Year to Dec. 2005*
All personnel	1.8	1.7	2.9	*	*	5.5
Officers	1.9	1.3	2.1	*	*	2.6
Navy	1.2	1.3	2.1	*	2.8	1.9
Army	1.5	0.6	1.5	*	2.9	3.4
Royal Air Force	3.1	2.9	2.9	*	*	2.0
Other ranks	1.8	1.8	2.9	*	6.9	5.8
Navy	1.6	1.4	1.9	*	3.3	3.5
Army	2.0	2.0	3.6	*	9.2	6.9
Royal Air Force	1.2	1.2	1.7	*	2.2	1.8

Source: DASA (Tri-Service).

Note
* Data unavailable.

trend. As a result of this improvement, a recent (23 February 2006) Ministry of Defence press release was able to claim:

> The percentage of UK Regular Forces from ethnic minority backgrounds continues to rise; at 1 January 2006 ethnic minorities accounted for 5.5 per cent compared to 5.2 per cent at 1 January 2005. The largest increase was in the Army, where ethnic minorities now comprise about 8 per cent of its regular forces.[14]

Despite this evidence of progress, however, it is necessary to enter some important caveats. The first is that, despite an upward trend since the publication of the SDR, latest figures suggest a fall-off in recruitment, although we should note that the data in the table cover different time periods. Closer inspection of the data suggests that there may have been a spurt in recruitment of minority ethnic personnel between 2000 and 2003, which may not necessarily be sustained.[15]

Second, there are important differences between service arms as well as between officer and other ranks recruitment. Greatest progress seems to have been made in the Army while in the Royal Navy and Royal Air Force (RAF) recruitment still lags behind target. Interestingly, it also appears that both the total numbers and the percentage of ethnic minority personnel entering the officer corps have declined relative to 2002 and that this decline has been most marked in the RAF.[16]

A third caveat concerns the conceptualization of 'ethnic minority' for the purposes of tracking progress. Although they are excluded from the figures presented in Table 10.1, we should note that Fijians, Vincencians and St Lucians, recruited *en bloc* to the Army, represented more than one-third of those recorded as members of ethnic minorities recruited in 2002/03. (This figure was itself down from half in the two previous years for which figures are available.) To this must be added Commonwealth citizens who have been recruited in the United Kingdom. This means that, despite evidence of more success in recruiting ethnic minorities in 2002/03, a substantial proportion of 'ethnic minority' personnel are, in fact, Commonwealth rather than British citizens. This distinction is important in the context of the history of differential and conditional incorporation during the imperial phase of Britain's past, not least because Commonwealth citizens are not entitled to the full suite of citizenship rights available to British citizens. This, in turn, raises two questions: does increasing diversity by these means really fulfil the spirit of the SDR commitments? And what implications does non-citizen status have for incorporation *on equal terms* within the military team?

Differential incorporation, voluntary exclusion and citizenship

The idea of representativeness embodied in SDR refers to a socio-demographic match between the military and society – one that can be attained through planned recruitment targets. The difficulty is that the gross category – 'ethnic minority' – takes no account of the different socio-demographic profiles, levels of social mobility, educational attainment and cultural traditions of the very diverse groups that make up Britain's ethnic minority population. It is entirely conceivable that the gross target of increasing ethnic minority representation to a level commensurate with the proportion of the population classified as belonging to an 'ethnic minority' could be reached without representativeness being achieved for some of the communities involved.[17] It is arguable that the progress that has been made in closing the gap in relation to the recruitment targets, reviewed above, exemplifies this problem.

Thus, in addition to the caveats entered in the last subsection, it is also important to note that the recruitment of indigenous minorities is significantly skewed towards those of Caribbean (and ultimately African) descent, including many with 'mixed' heritage. South Asian-descended personnel remain under-represented in the armed services to the point of near invisibility. This under-representation of South Asians also raises questions about the extent to which, even with improved performance against the SDR targets, the armed forces are really more representative of British society. It suggests that the new-found diversity is, in practice, limited in scope.

This may be the result of inadequate or mistargeted recruitment efforts. It may, however, also be the result of self-selection – that is, an unwillingness to enlist. The armed services have long tended to see the barriers to improving the recruitment of ethnic minority personnel as lying within ethnic minority communities – in ignorance about the opportunities afforded by the services or from concern about a past racism that is now said to be being addressed. There is some evidence that there is an element of truth in these explanations. But there are also some indications of a general sense that the armed forces do not really 'belong' to ethnic minority citizens and that this also explains the apparent reluctance of some groups to enlist.[18]

We should note, however, that there are dangers in the tendency to read voluntary exclusion in cultural terms. Importantly in this connection, we should note that there may be a differential propensity for members of different groups to select particular occupations or to aspire to particular careers. Moreover, we know that the 'white' population has not had a uniform propensity to select the armed forces as a career and we should not be surprised if similar differentials were to be found among other groups. Indeed, we know that most whites do not enlist and that those who do tend to be drawn from highly traditional, and geographically discrete, recruitment pools.[19]

This suggests that cultural preferences are likely to play only a part in decisions about whether or not to join the armed services. The absence of a tradition of service and, crucially, of vicarious experience mediated through family or friendship networks may be quite critical. In addition, we should not underestimate the role of rational decision-making.

Of some importance in this context is the growth of a middle class of professional and managerial workers in some ethnic minority communities. Indeed, some have suggested that there is a developing convergence in the class structures of ethnic minority groups towards that of the majority white population.[20] The most recent analyses of data for the late 1990s suggest that in terms of economic activity, unemployment and job levels, even some of the previously most disadvantaged groups, namely Pakistanis and Bangladeshis, have continued to close the gap with whites.[21]

There remain, of course, significant differences between members of different ethnic groups, with those of Caribbean descent in particular apparently still lagging behind. Moreover, there are marked regional variations, with significant local pockets of labour market disadvantage and exclusion affecting even those groups apparently experiencing upward mobility. We should also remember that upward mobility is quite compatible with occupational segregation and continuing discrimination.

Nevertheless, this evidence should alert us to the need to place any discussion of the attractiveness to potential ethnic minority recruits of a career in the armed services in the context of a recognition that the range of other opportunities open to them is also changing. The evidence on educational attainment further reinforces this point. It is clear that members of ethnic minority groups are more likely to remain in full-time education after the age of 16 than are their white counterparts – a finding that holds for both young men and young women.[22] Moreover, the evidence suggests that this is a relatively long-standing pattern dating back at least to the beginning of the 1980s.[23] In addition, it appears that ethnic minority groups are generally over-represented in higher education relative to their presence in the population as a whole, although this gross observation conceals some important variations.[24] In this connection, Modood has argued that it is important not to underestimate the strength of what he calls 'ethnic minorities' drive for qualifications', which he attributes to a strong motivation for economic betterment.[25]

Any analysis of the potential attractiveness of a military career must be placed in this broader context. Indeed, some recent survey evidence about the attitudes of minorities of Pakistani Muslim descent in Britain lends weight to this line of argument. When questioned about the attractiveness of a military career, respondents in the survey cited concerns about racism as the major issue accounting for their lack of interest. Significantly, however, the authors also found evidence of more generalized concerns about the nature of a military career as well as a tendency to prioritize further and higher education over enlistment in what was often seen as a low-status occupation.[26]

Looked at from this point of view, voluntary exclusion may represent as much the outcome of rational choice-making as of culturally framed preferences. The question is less one of choosing not to serve than one of prioritizing other opportunities, such as the independent professions, over the option of a military career. In this connection we should note that research into the apparent unwillingness of members of Britain's South Asian minority ethnic groups to enter the nursing profession found similar patterns in that there appeared to be no shortage of applicants from these same groups for careers as doctors.[27]

If this is correct, then the under-representation of some groups in Britain's

Britain 147

armed forces may represent less the denial of citizenship and more a rational choice not to participate – that is, voluntary exclusion. But this does raise the question of the duties as well as the rights of citizenship. What level of voluntary exclusion, we might ask, is consistent with claims to citizenship rights? Posing the question in this way, however, again problematizes minority cultures. Does the same question not also apply to those white people (the majority) who reject military service?[28]

Building on success

The challenge of increased numbers

As we saw at the outset, however, the citizenship case is not the only one underpinning the drive for the improved recruitment of minority ethnic groups. Business needs, in the context of continuing recruitment challenges, mean that success in meeting the SDR and subsequent targets has potential operational benefits. Even given the caveats set out above, it is clear that, for whatever reasons, there has been a growth in the numbers of minority personnel. This in turn, however, poses its own challenges.

For the first time, retention in particular becomes a key issue in maintaining diversity. Indeed, it is arguable that recruitment and retention are inextricably linked and that the barriers to change are located not only before and at the point of entry but also within the organization itself. In this respect, internal cultural change is one of the keys to the success of recruitment initiatives. Without this, the recruitment of increasing numbers of ethnic minority service personnel could magnify the negative image of the forces among ethnic minority communities if the experience of recruits were to be that little has changed. In other words, recruitment-centred initiatives could be dangerously self-defeating unless internal change is also seriously addressed. There are a number of dimensions to this.

One concerns advancement. As research on gender equality has shown, this can quickly replace recruitment as the key issue in ensuring equity. Is there a danger that recruitment barriers will be replaced by an (armoured) glass ceiling with the NCO corps representing the effective upper ceiling for ethnic minority personnel? Here differential recruitment routes to commissioned and non-commissioned ranks may be of great significance, particularly given the data adduced earlier relating to differential recruitment to these routes. Were this to be the case, it might well reinforce the image of the armed services as a low-status profession, with the *perceived* exclusivity of the officer corps operating as its own disincentive. Does this suggest that a truly radical review might problematize the distinction between commissioned and non-commissioned personnel altogether? What operational advantages are conferred, notably in modern, high-tech military environments, by a distinction that has its origins in the class structures of medieval Europe?

A second question concerns the role that occupational cultures can play in the success of equal opportunities or diversity initiatives. Their significance relates first to the direct impact on compliance with the requirements of equal opportunities policies, recruitment practices and advancement policies. In addition, there is evidence that they contribute indirectly through the experiences of ethnic minority personnel. Here the work of Holdaway and Barron on the police service is particularly instructive.[29] Their data show, for example, that the ability of ethnic minority personnel to fit into the organization is directly related to their capacity to participate in the occupational culture of the police and the extent to which they are allowed to do so. Their willingness to do so is, at the same time, a key to their acceptance by other officers. There is, in other words, a potentially vicious cycle of exclusion into which it may be difficult to break. The problems are particularly acute where the occupational culture is centred on activities from which some personnel are excluded. Muslim personnel, for example, may feel unable to participate in activities centred on the consumption of alcohol, while their lack of participation may be seen by other officers as an unwillingness to commit to the team. It is not difficult to see how such processes might have parallels in military organizations.

A third issue concerns equitable treatment. Much of the high-profile negative publicity suffered by the British armed services has concerned racial harassment as an aspect of a wider problem of bullying, examples of which continue to surface.[30] In building on the successful increase in recruitment, it will be important to guard against the possibility that greater numbers of ethnic minority personnel could result in more high-profile cases and bad publicity, with obvious knock-on effects for recruitment.

Fourth, there is the danger of conditional inclusion. We noted above that some of the apparent success in meeting recruitment targets results from a reliance on Commonwealth – as opposed to full British – citizens as a solution to the recruitment dilemma. Is it possible that this could lead to a situation in which some comrades are seen as more equal than others – mercenaries rather than fellow citizens? Here we should not underestimate the importance of Britain's military history. Minorities have not always been excluded from the British armed services. Indeed, the policing and defence of Empire led to the recruitment of large numbers of imperial subjects to the armed forces. Large numbers of such personnel served Britain in both world wars – a fact celebrated in a recent (recruitment-oriented) Ministry of Defence touring exhibition entitled 'We Were There'.[31] However, the Empire's inclusion of such personnel always was highly conditional. They were most often recruited to separate units officered by British personnel (as in the earlier case of the British subnational minorities: Scots, Welsh and Irish). The continuing role of Gurkha regiments recruited from Nepal into the British Army remains an obvious case in point.[32] The 'We Were There' message could, therefore, as easily stir painful memories of imperial subjugation and exploitation as of pride in a shared military past.

Embracing diversity

All of this suggests that building on success is not merely a matter of technical operational changes in the arrangements for recruitment, selection and advancement. It is about rising to the fundamental challenge of embracing diversity. In a military context, diversity has to mean more than simply accommodating different cultures, skin colours or traditions. It must entail a valuing of diversity within a shared framework of values and expectations. Difference must be accommodated to the need for uniformity, discipline, *esprit de corps*, comradeship and cohesion that are core components of military self-image and organization. Their most visible manifestation is the uniform, which de-emphasizes individuality and emphasizes similarity, sharing, and the regulation of the group by a distinct body of norms and laws. Can we be confident that British understandings of diversity have yet reached this level of sophistication?

In this respect, the heritage of Empire continues to weigh heavily on the British national psyche. Its loss has not yet been fully come to terms with – a fact that manifests itself in everything from everyday racism to the near-universal expressions of support for 'our boys', even in the context of the otherwise unpopular war in Iraq. Moreover, the relationship between citizenship and nationality poses particular problems of its own because of the way in which both Englishness and Britishness are routinely represented as uniquely long-standing and primordial attachments.[33]

The significance of this appeal to historical continuity is greatly heightened when we consider the fact that all armed forces tend to place a very high value on tradition and history. In the case of the British armed forces this emphasis on history has a particular significance, since much of the military history of Britain over the past two centuries or so is the history of colonial involvement. Many of the campaigns fought by the British armed forces were either against colonized peoples or to protect imperial territory from other colonial powers. Thus, the recent forebears of many of Britain's citizens who are not white were either enemies or colonial subjects. In these circumstances it may be difficult to view their descendants as co-nationals – whatever their formal citizenship – because they lack both the common origins and the ethnic homogeneity which the British national myth, with its claims to a uniquely long history, requires. In this respect, British self-conceptions diverge markedly from those of the United States, whose origin myth emphasizes diversity and whose official national values centre on a contractually devised Constitution.

Practical challenges

Against this background, what are the challenges that must be faced up to? First, it is necessary to build on recruitment success by ensuring good retention, providing opportunities for advancement, ensuring equity of treatment and, above all, making certain that all publicity is, indeed, good publicity. Second, there is a need to sell the benefits of diversity. This means making the most of diverse talents in a military ever more required to exercise flexibility and initiative in the face of increasingly diverse missions. Personnel with diverse individual and collective experiences can make a major contribution, and their contribution needs to be highlighted and valued. However, this increasingly varied operational mission poses its own problems.

What are the implications of a lack of clarity about the boundaries between liberation and conquest, or between peacekeeping, occupation and war-fighting – particularly for personnel who may have more or less tenuous connections with local populations? At what point does diversity turn from a benefit to a liability, and under what conditions? This places great significance on the need for clarity about the political objectives of military action. It also means having the vision to distinguish real from imagined conflicts of loyalty.

Post 9/11, it is commonplace to see Islam as providing a particular challenge in this context. Many of the diagnoses fall victim to a tendency to essentialize Islamic communities, ignoring national, communal and doctrinal differences (for example, between Turks in the Netherlands and Pakistanis in the United Kingdom). In practice, there is no strong evidence either that Muslim personnel are especially victim to divided loyalties or that in other circumstances other groups would not experience tensions of this kind. Indeed, evidence from the 2001 Census suggests that the vast majority of all those affirming membership of an ethnic minority group also unequivocally described their national identity as British, English, Scottish or Welsh. This included almost nine in ten people from a 'Mixed' (88 per cent) or 'Black Caribbean' (86 per cent) group, around eight in ten people from a 'Pakistani' (83 per cent), 'Bangladeshi' (82 per cent) or 'Other Black' (83 per cent) group, and three-quarters (75 per cent) of the 'Indian' group.[34]

Yet the persistence of speculation about this issue, taken together with a popular tendency, in Britain at least, to demonize all Muslims as potentially fundamentalists or terrorists does have potential consequences for unit cohesion, the goal of which is, of course, the tested solution to problems of loyalty, commitment and, crucially, military effectiveness. Evidence from the United States suggests that one of the biggest challenges to unit cohesion is unresolved racism and bullying.

In conclusion, we would suggest that there are a number of keys to ensuring that racial and ethnic tensions do not undermine the cohesion that is at the heart of all military activities. The first is building a commitment to valuing diversity from the political leadership down through the chain of command. The British equal opportunities directive is a key aspect of this process, but it must be underpinned by real implementation at an operational level in the everyday conduct of military life.

This means, second, that the operation of the chain of command is central to the effective delivery of the armed forces' commitments. Evidence from private-sector equal opportunities initiatives[35] suggests that what is required is a long-term strategy with commitment and leadership from the top. Securing the commitment of lower line managers and supervisory staff is frequently the key to success. In addition, it implies careful attention to key break points in the system for 'cascading' policy to ensure that the day-to-day operational demands made on lower line managers do not lead to diversity having a low priority for them. Where such individuals also play a key role in shaping the occupational culture, this may ensure that even board-level commitments and strategies are sidelined or circumvented. In the armed forces, key personnel in this regard are likely to be non-commissioned officers. This means that the beating heart of the military machine – the NCO corps – must be fully engaged and committed to the diversity agenda. This, in turn, means that they must be fully convinced of its operational benefits as well as of its moral rectitude and its congruence with wider societal values.

Herein lies the greatest challenge. The armed services are the servants and representatives of a wider society, doing its bidding, reflecting its traditions and defending its values. But this raises the question of what it is they are to be representative (or reflective) of. The challenge of securing the inclusion of ethnic minority citizens – of creating the conditions in which diversity can be valued in all aspects of social life – is a national one, not just a task for the armed forces. In Britain this entails nothing less than completing the task of re-imagining the nation in the twenty-first century.

Notes

1 In this chapter we refer both to the United Kingdom and to Britain. Technically, the armed forces with which we are concerned are those of the United Kingdom, which includes Britain (that is, England and Wales, and Scotland) and Northern Ireland. In practice, the relationship between Britain and Northern Ireland is a complex and contested one. In much popular discussion, as well as in government publications to which we refer, the term 'Britain' is more commonly used. Moreover, for the purposes of some parts of our argument, references to Britain are technically the correct ones. Similarly, in our references to imperial and colonial history, Britain is the appropriate unit of analysis. We recognize that these usages may give rise to some inconsistencies, but these cannot be adequately resolved within the confines of this chapter.
2 *The Strategic Defence Review*, Cm 3999, July 1998.
3 Ibid., Supporting Essay 9, para. 7.
4 See BMSG, *The Future of British Military Cultures*, October 1997.
5 www.mod.uk/DefenceInternet/AboutDefence/Organization/KeyFactsAboutDefence/ DiversityAndEqualityITheArmedForces.htm (accessed 4 April 2006).
6 Commission for Racial Equality, *Report of a Formal Investigation into the Ministry of Defence (Household Cavalry)*, 1996.
7 2. *The Strategic Defence Review*, Cm 3999, July 1998, Supporting Essay 9, para. 41.
8 Note that the population figure in this statement has been superseded by the results of the 2001 Census, which show that some 9 per cent of the population of England affirmed ethnic minority group membership. The target for ethnic minority recruitment has also been raised from an initial aim of 6 per cent by 2006 to 8 per cent by 2013.   www.mod.uk/DefenceInternet/AboutDefence/Organization/KeyFactsAbout Defence/DiversityAndEqualityInTheArmedForces.htm (accessed 4 April 2006).
9 See C. Dandeker and F. Paton, *The Military and Social Change: A Personnel Strategy for the British Armed Forces*, London: Centre for Defence Studies/Brassey's, 1997.
10 The work of Lt Col. Stuart Crawford is of particular note in this connection. The arguments developed here draw on Crawford's work and on a number of conversations with him. See S.R. Crawford, 'Racial Integration in the Army: An Historical Perspective', *British Army Review*, 111, December 1995, 24–28.
11 According to the 2001 Census, the figures are 7.9 per cent of the population of the

United Kingdom and 9.0 per cent of the population of England. www.statistics.gov.uk/cci/nugget.asp?id=455 and www.statistics.gov.uk/cci/nugget.asp?id=457 (accessed 4 April 2006).
12 www.statistics.gov.uk/cci/nugget.asp?id=456 (accessed 4 April 2006).
13 See T. Modood, R Berthoud, J. Lakey, J. Nazroo, P. Smith, S. Virdee and S. Beishon, *Ethnic Minorities in Britain*, London: Policy Studies Institute, 1997,
14 www.mod.uk/DefenceInternet/DefenceNews/PeopleInDefence/UkArmedForcesAt983PercentManning.htm (accessed 4 April 2006).
15 It is also possible, of course, that the war in Iraq may have had a negative effect on the recruitment of ethnic minority personnel.
16 www.dasa.mod.uk/natstats/tsp4/tsp4tab3.html (accessed 4 April 2006).
17 Research has demonstrated patterns of this kind in the nursing workforce, for example. See P. Iganski and D. Mason, *Ethnicity, Equality of Opportunity and the British National Health Service*, Aldershot, UK: Ashgate Publications, 2002.
18 See A. Hussain and M. Ishaq, 'British Pakistani Muslims' Perceptions of the Armed Forces', *Armed Forces and Society*, 28, 4, 2002, 601–618.
19 See, for example, C. Dandeker and A. Strachan, 'Soldier Recruitment to the British Army: A Spatial and Social Methodology for Analysis and Monitoring', *Armed Forces and Society*, 19, 2, Winter 1993, 279–290, and H. Strachan, 'Reassessing Recruitment Strategies for the Armed Services', in A. Alexandrou, R. Bartle and R. Holmes (eds), *New People Strategies for the British Armed Forces*, London: Frank Cass, 2002, pp. 100–115.
20 P. Iganski and G. Payne, 'Declining Racial Disadvantage in the British Labour Market', *Ethnic and Racial Studies*, 19, 1, 1996, 113–134; P. Iganski and G. Payne, 'Socio-economic Re-structuring and Employment: The Case of Minority Ethnic Groups', *British Journal of Sociology*, 50, 2, 1999, 195–216.
21 P. Iganski and G. Payne, 'Social Exclusion and Social Inclusion: Britain's Minority Ethnic Groups', paper presented to the Cambridge Stratification Seminar, Claire College, Cambridge, 19–21 September 2000; D. Mason, 'Changing Ethnic Disadvantage: An Overview', in D. Mason (ed.), *Explaining Ethnic Differences: Changing Patterns of Disadvantage in Britain*, Bristol: Policy Press, 2003; D. Mason, 'Changing Patterns of Ethnic Disadvantage in Employment', in Mason (ed.), *Explaining Ethnic Differences*.
22 Modood et al., *Ethnic Minorities in Britain*.
23 C. Brown, *Black and White Britain: The Third PSI Survey*, Aldershot, UK: Gower, 1984; UK Department of Education and Science, *Education for All: The Report of a Committee of Inquiry in the Education of Children from Ethnic Minority Groups*, Cmnd 9453, London: HMSO, 1985; D. Drew, *'Race', Education and Work: The Statistics of Inequality*, Aldershot, UK: Avebury, 1992; T. Jones, *Britain's Ethnic Minorities*, London: Policy Studies Institute, 1996.
24 Jones, *Britain's Ethnic Minorities*, p. 32; Modood et al., *Ethnic Minorities in Britain*; T. Modood and T. Ackland, *Race and Higher Education*, London: Policy Studies Institute, 1998; T. Modood and M. Shiner, *Ethnic Minorities and Higher Education: Why Are There Differential Rates of Admission?*, London: Policy Studies Institute, 1994.
25 T. Modood, 'Ethnic minorities' drive for qualifications', in Modood and Ackland, *Race and Higher Education*.
26 Hussain and Ishaq, 'British Pakistani Muslims' Perceptions'.
27 Iganski and Mason, *Ethnicity*.
28 Charles Moskos makes a related point in his critique of non-elite participation in military service in the United States. See his debate with Korb in www.taemag.com/docLib/20030117_views1201.pdf.
29 S. Holdaway and A. Barron, *Resigners? The Experience of Black and Asian Police Officers*, Basingstoke, UK: Macmillan, 1997.

30 See, for example, the inquiry into the deaths of four recruits at the Deepcut army barracks. news.bbc.co.uk/1/hi/uk/4856148.stm (accessed 4 April 2004).
31 www.wewerethere.mod.uk/intro.html (accessed 4 April 2006).
32 See, for example, T. Gould, *Imperial Warriors: Britain and the Gurkhas*, London: Granta, 1999.
33 L. Colley, *Britons: Forging the Nation, 1701–183,* New Haven, CT: Yale University Press, 1992; P. Rich, *Prospero's Return: Historical Essays on Race, Culture and British Society*, London: Hansib Publishing, 1994.
34 www.statistics.gov.uk/cci/nugget.asp?id=459 (accessed 4 April 2006).
35 See N. Jewson, D. Mason, A. Drewett and W. Rossiter, *Formal Equal Opportunities Policies and Employment Best Practice*, Research Series no. 69, London: Department for Education and Employment, 1995.

11 Diversity in the French armed forces

Bernard Boëne and Claude Weber

France's military establishment has been labouring under an all-volunteer format for the past four years – in a context marked by what has variously been termed 'radical modernity' or 'post-modernism', where multicultural diversity figures prominently. The time seems ripe for an early assessment of the trends such a major change has brought in its wake. Analysis of the issues relating to the management of diversity provides, for that purpose, a vantage point of choice.

Where we come from: diversity in conscription's final decade

When the last conscripts left the armed services in late 2001, a mixed format of mostly professional cadres and drafted rank and file came to an end after nearly a century of existence. In its closing decade, dealing with the topic of ethnic, gender or cultural diversity proved a rather dull exercise. The low visibility of minorities at all levels, combined with the usual effects of brotherhood in arms, had until then been reason enough for the issue to remain dormant. The absence of controversy surrounding women in the forces, or the fact that gays and lesbians in their ranks were a non-issue, explained the rest.

Long-term societal trends

The same could long be said of the country as a whole. Feminism and homosexual militancy had remained less strongly expressed in France than in most comparable countries. To all appearances, until the early 1980s the rising tide of immigration generated by rapid economic growth and the industrial workforce requirements it entailed had raised few, if any, problems. True, the decision made in 1974 to allow foreign immigrants to bring their families over, combined with an initially liberal naturalization policy, had changed the equation subtantially over time as the sons and daughters of immigrants proved much less docile than their parents in reacting to discrimination. But the welfare treatment applied to the throes of a nascent multicultural society – protest marches, even riots – seemed to work, and for a while France lived under the delusion that adjustment would require little extra effort. The triumph in the 1998 soccer World Cup of a 'rainbow' soccer team, and the enthusiasm it elicited in nearly all segments of the population, were widely hailed as living proof that a multicultural, multi-ethnic society was well on its way to success.

In many ways, such a benign (though in retrospect, fragile) condition was the legacy of a long-standing national republican tradition in which de-emphasis of internal differences, a rather sharp delineation of public and private spheres, and cultural assimilation of newcomers played a crucial part. Until the late 1970s, French society was a pretty tightly integrated polity in which powerful social forces ensured that minorities of foreign (mostly Italian, Polish, Spanish, Portuguese) workers and their families were soon turned into citizens. While women did not enjoy political citizenship rights until 1945, and legally remained subject to male domination until thirty years later, the general tone of gender relations, still marked by the legacy of *Ancien Régime* court society, was less adversarial than in English-speaking countries.[1] Gays, until the 1990s, rarely came out of the closet, but then sexual orientation was (and for the most part still is) regarded as a strictly private matter and, with the possible exception of the norms prevailing among the working classes, a widespread climate of tolerance had come to remove much of the sting of social stigma formerly associated with homosexuality.

However, optimism was mitigated by the realization that the so-called republican model built, over a century earlier, around formal equality of opportunity, social solidarity and political purpose, internal (emancipation of citizens) and external (France as beacon of liberty to the world), was showing serious signs of wear. In an age of rising individualism and weakening citizenship norms, immigration – much of it Islamic in origin – had for some time begun to exceed the country's capacity for assimilation, and led to a new process of accommodation that was not exactly painless.[2] Bearing witness to a novel situation was the frequency of serious riots in the depressed suburban areas where, owing to poverty, white flight, discrimination and lack of meaningful integration to the wider society, families of immigrant origins now were concentrated to such an extent that it has become relevant to speak of segregation.

One wing of the feminist movement, supported by moderate women exasperated by their under-representation in responsible positions in most sectors of society, suddenly demanded parity in access to elected or appointed political office. As for homosexuals, the past few years had seen activist associations and gay pride marches alter mainstream perceptions to a rather surprising extent. Practices hitherto unheard-of, such as 'outing' public figures said to be gays, or the announcement by a leading female tennis player that she is a lesbian, had recently become part of the French scene. In brief, change was on the cards. In a culture long premised on minimizing differences in the public sphere, and a society or organizations variously described by classic social scientists as stratified by status honour[3] and characterized by the avoidance of face-to-face relationships,[4] such trends, imported or indigenous, were bound to create problems.

The issue ran even deeper, as these emerging trends appeared to be part of a wider pattern. In advanced democracies, at once egalitarian and libertarian, 'post-modern' tendencies, based on a fragmentation of personality, culture and society, had become apparent as from the 1980s. Identities, increasingly chosen (and changed) rather than ascribed, now defined shifting communities cultivating values, lifestyles and agendas of their own. As a result, societal norms were increasingly seen as forms of majoritarian tyranny. In some quarters, personal and communal rights started to enjoy pride of place at the expense of citizens' duties. In such a societal state of affairs, because they favour (but hardly guarantee) relatively peaceful interactions, the only normative values that survive the loss of cultural integration are tolerance, equal dignity of recognized identities, and the sanctity of life. While France seemed less distinctly affected by such tendencies than comparable countries of North America or Europe in the same time bracket, it too was caught in a slow yet powerful process that has seen it gradually moving from a culturally homogeneous, socially differentiated configuration to one in which politics is increasingly governed by cultural diversity rather than class affiliation.

The military in the 1990s

Yet diversity and post-modernism were then too novel to have influenced the military establishment. Conscription, which up to the end mainly affected white males drawn from the secularized Christian majority, ensured that manpower requirements were met, and provided no incentive to call in question the forces' cultural conservatism. Much of the pulling and hauling that occupied centre stage in countries where culture wars resulted from diversity and its problems was absent from the French military scene: women, gays or members of ethnic or racial minorities in uniform still had to create a stir.

Minorities

True, after one generation and more, France now woke up with sizeable ethnic/racial groups of military age. The exact numbers were (and still are) difficult to gauge, since a constitutional principle forbids any mention of race in identification documents or census data. But the military had long had little difficulty in addressing such issues.

For one thing, minorities were under-represented within its ranks for a variety of reasons. Among those of conscript service age, many found themselves exempted in practice, for example because they helped support their families or they did not pass aptitude tests, or again because they held dual citizenship, and bilateral agreements between France and the other nation involved (e.g. Algeria) enabled them to escape service under arms in the country of residence. Some simply failed to register. For another, immigrants do not generally move to France in order to serve in the armed forces. Some of their sons did, however: in the 1980s and 1990s, the combat companies of the few all-volunteer battalions earmarked for intervention overseas comprised minorities whose proportions were known to exceed estimated national proportions by a factor of two or three.

Those who did enlist came to the military mainly to acquire 'first-class citizenship', or to escape anomic conditions in their families or social milieus. Minorities, however, were still hardly visible in the officer corps.

Another reason for the few problems raised by minorities is that colour feeling in French society had until recently been distinctly lower than in many other Western countries. But the main factor resides in the capacity of military institutions to use shared experiences, the hardships of service in the field, and the fraternity of arms in order to integrate service members of diverse origins. The Foreign Legion's reputation in that regard is by no means undeserved. Nor is this phenomenon purely French: the US military has since the early 1950s succeeded much better than has its parent society in solving the race relations problem.[5]

Servicewomen

Although integration of females had been less extensive than in the US military, by European standards the French armed forces had not lagged. As recently as the mid-1970s, military women had to struggle individually for acceptance and recognition in a world obviously not meant for them. In the 1990s, with numbers stabilized at slightly above 20,000 (less than 5 per cent overall, 7.5 per cent of career and contract personnel), nobody seemed to notice or mind their presence, though in relative terms it was increasing in proportion to the decline in total force levels. Such progress had been achieved gradually, without a hint of controversy. This was related in part to the fact that personnel managers appreciated women as a source of high-quality recruits: the rate of selection, which often hardly exceeded one in two or three among men, frequently rose as high as one in ten or one in fifteen for military women because of low regulatory ceilings.

Female service members – all of them volunteers under the mixed format – had, in theory, been almost fully integrated since the early 1980s; they were no longer limited to 'traditional' jobs or to the lower levels of the hierarchy. Women represented over 8 per cent of NCOs and volunteer other ranks. Because the younger generation among them had been granted access to service academies in 1983, their representation and prospects in the officer corps were slowly improving. They gradually penetrated male fortresses in increasing numbers. Even in the tradition-oriented Navy, since 1992 they had been authorized to serve on board, and one was a ship commander. The opening, in 1996, of fighter pilot slots to female Air Force Academy graduates made them more central symbolically. Finally, with only a handful of exceptions, all regulatory quotas were abolished in 1998.

In practice, however, almost three-quarters of military women were sergeants engaged in specialist jobs, and few female service members volunteered for positions in the combat arms. Only a tiny minority were willing to quarrel with this situation. Dissatisfaction was hardly evident, if only because (as surveys showed) women's reasons for joining the services were job security, a highly structured environment, a military family background, and the desire to do something different from the routine jobs in which civilian women tend to concentrate. External pressure for further advances was lacking, as, put off by servicewomen's conservative orientations, feminist movements in France, characterized (unlike their American counterparts) by fairly strong anti-military feeling, neglected the armed forces as a possible symbolic battleground. French military women were thus a prime example of emancipation achieved through non-adversarial gender relationships.

Homosexuals

In the past, military and civilian society did not differ much with regard to same-sex relationships. France was a highly normative society, but because of its Catholic sensibilities it tolerated deviant behaviour at the margins, if only because it could be pardoned in confession. In addition, the idea that one must live in a glass house is alien to the French cultural tradition. This explains why one finds great military figures with known (though not open) homosexual tendencies, such as Marshal Lyautey, or even why, as illustrated in all sorts of legends, homosexuality was proverbial among colonial troops.

In the 1990s, although homosexuality was now recognized as 'normal' in civilian society, the issue was rarely raised in the military. Indeed, it was shrouded in the kind of silence that signifies neither embarrassment nor a sense of taboo, but complete lack of interest. Out of 15,000 registered disciplinary incidents in 1992, only twenty were related to sexual matters; only five of these involved homosexuals.[6] Part of the reason, again, was cultural: sexuality belongs to the realm of inviolable individual privacy. Nobody is required to declare his or her sexual orientation, or told that homosexuality is incompatible with military life. The norm is that sexual activity is off limits in military precincts. There was thus little incentive for gays to vindicate their sexual conduct publicly. Furthermore, it is likely that gays were under-represented in the services, as self-described homosexual conscripts were often discharged informally by service doctors before induction, on the unproven but not entirely unreasonable grounds that they might create problems of cohesiveness in their units. The gays who did serve, self-selected or otherwise, must have been fairly conservative, because very few evinced a burning desire to stir things up.

Scrutinizing recent change: a set of hypotheses

Analysis of the change that has affected the French military in the past few years requires that attention be directed to the impact of its main driver – the advent of an all-volunteer format – and the context in which it has taken place. It also requires hypotheses, derived from the observation of empirical trends in comparable countries with a longer history of all-volunteer forces (the United States, the United Kingdom), as well as from recent international social science literature on these topics.

Expected impact of the new format

The historic switch to an all-volunteer force is bound to have changed the equation substantially. Women are destined to become much more visible. Their numbers will rise as male enlisted accessions become vulnerable to shortages. In the context of a rapid reduction in force by over 30 per cent, their overall proportion is likely to increase mechanically well beyond the 10 per cent mark. Having learned that it cannot do without them if it is to fill the ranks at all times with quality personnel, the military will further accommodate their needs and aspirations. The same is likely to apply to ethnic and racial minorities when the third generation among the earliest waves of immigrants come of military age (by which time the proportion of non-whites and/or people other than secularized Christians will in all likelihood start to rise in the ranks).

Discrimination will raise problems, and will have to be eliminated for reasons of cohesiveness and political legitimacy. The military will even be invited to act as a vehicle for effective integration into the larger society. However slightly and slowly, cultural conservatism will weaken. Gay and lesbian service members will probably benefit from the new situation – in the subdued ways that befit the French cultural tradition. Post-modern trends, though they partly point in different directions, will do the rest. It does not take great analytical skills to predict that managing diversity will at some point become an issue.

Context: the rise of post-modern standards of representativeness?

Under post-modern conditions, individuals choose from among a number of potential identities one that will provide their lives with expressive meaning. Such identities often accentuate traits derived from psychobiological tendencies (e.g. gays and lesbians) or objective characteristics (race, gender), but do not necessarily do so. Some are not grounded in any traits that formerly would have provided the basis for ascribed status, but rather express personal interests, tastes or philosophies. The anecdote about three Italian-American female jazz buffs who decided they would live as African-American women attests to the fact that what was once regarded as objective is increasingly coming, at least in part, to be defined subjectively.

In such a configuration, the need for representativeness then acquires new meaning. The novelty resides in the fact that public institutions increasingly rely for their legitimacy on *cultural* rather than on *social* diversity. Personal choice in cultivating identities will further undermine integration and citizenship norms, then seen as an unnecessary mutilation of personality and freedom. Representativeness, not so long ago, was assessed in a strict mathematical, i.e. proportional, sense (though the multiplication of criteria – socio-economic background, race, gender, ethnicity, etc. – complicated the issue to the point of rendering it operationally unmanageable). The entity which served as baseline for measuring it was the polity as a whole (an ideal easier to approach under draft conditions than in all-volunteer formats, because the latter tend to reflect the compositional make-up of the working population rather than the citizenry). Under-representation of a given group, where it existed, was the product either of a low propensity to join the forces among its members, or of institutional discrimination against it. Over-representation was the mechanical effect of under-representation among other groups, or the expression of a higher propensity to serve under arms – as the case may be due to a disadvantaged group's strategy purporting, through merit, to acquire first-class citizenship in a society that denied it (e.g. blacks in the United States from the early 1950s onwards).

Both under- and over-representation carried penalties for the military from a functional as well as sociopolitical standpoint. In terms of recruitment and retention, they were apt to trigger vicious circles: suspicion of discrimination against a given underprivileged minority would depress its propensity to enlist, while its over-representation, if it reached significant proportions, entailed a risk of recruitment shortfalls among mainstream groups, whose potential candidates for enlistment would then shy away from the military. Socio-politically, the penalty was in terms of a less than satisfactory public image of the armed forces as an unrepresentative institution, possibly leading in turn to a less than optimal amount of public support. Remedial policies revolved around integration through equal opportunity.

The new imperative of cultural diversity seems to raise fewer problems in that regard. Since society then turns into a mosaic of recognized cultural minorities, and the armed forces into a tribe among tribes, allowed to cultivate its own functional values and lifestyle, the issue of representativeness is easier to solve. Put differently, under conditions of free choice, low force levels and cultural diversity, representativeness of the military becomes qualitative and symbolic rather than mathematical. Under-representation is less of a problem.[7]

Future problems in managing diversity

If that is the case – or, more precisely, if emerging post-modern patterns one day become dominant – then quite a few difficulties can be anticipated. One is that the transition period is likely to last for some time. The mainstream of societies will become fragmented, and identities shift from ascribed to chosen, but they will not do so overnight or even in a matter of a few years. Such being the likely case, many individuals will continue to play by the old rules of integration and de-emphasis of differences in the public sphere. Not all gays and lesbians wish to advertise their sexual preference; not all feminists or members of ethnic or racial minorities want to opt for a separate lifestyle. Such a dual-track situation will probably be hard to manage, because policy will as a result be torn between conflicting principles. The policies applicable in either case, however, do have an important element in common: they share the need for a level playing field.

Another difficulty is raised by living in close confines. Unease may result from clashes of expectations, practical realities, relative deprivation or, in the worst case, harassment. The recent American controversy as to whether men and women should undergo basic training separately[8] (as in the Marine Corps) or not (as in the Army, Navy and Air Force), and whether physical criteria should be the same regardless of gender, is a classic illustration of that difficulty. And there is, of course, the whole issue of bonding and cohesiveness: does diversity weaken cohesion at both unit and secondary levels?[9] Beyond that, armed service still is, though admittedly in looser fashion under an all-volunteer force than under a draft, connected to national citizenship, which in a democracy implies integration.

Remedies and their related difficulties

There is only so much that detailed regulations, reaffirmation of ideals, and resort to affirmative action can do in the face of a mounting tide pointing in a different direction. What, then, is to be done? A solution is suggested by contemporary US university campuses, where dormitories have become (at least partly) segregated again. This has come about for the same reasons it might one day in the military: the demand for recognition by identity groups with a political agenda, and the need to avoid politically sensitive friction or harassment. Repeated, well-publicized instances of sexual harassment of military women in the United States recently illustrated the point. Such a solution would involve mixed units made up of homogeneous subunits, thus reaping the functional benefits of homogeneity in terms of cohesiveness. Effectiveness *within units* would, incidentally, also be enhanced by a further factor, usually to be seen at play *among* units: competition or emulation.

This is by no means a perfect remedy. Post-modern affiliation to an identity group being subjective rather than objective, there will be those, as already mentioned, who choose not to stress their gender, race, ethnicity or cultural references, and continue to value integration in a polity governed by national political purpose rather than by community feeling. These service members will have to be accommodated in 'rainbow' units of the sort we have grown accustomed to in the past decades, and which will probably remain the majority for some time. As Foreign Legion officers know full well, while primary and secondary cohesion may be more difficult to achieve in units and small groups that are culturally diverse, it is not altogether impossible: it merely takes longer to promote, other things being equal.

Predictions

The final hypothesis bears on the degree to which the above conjectures, made plausible by similar trends observed elsewhere as analysed in the literature, will be borne out by the data derived from the most recent studies available on the French case.[10] It may well be that the prism of French cultural tradition will serve to blunt their impact on the military, and beyond on society. It is likely also that France is less advanced than comparable countries as regards the incidence of post-modern trends. Issues relating to gender and sexual preference will probably be felt to be less acute in the French context than in others, as indeed has been the case until now. But in the longer term, racial/religious minorities may choose to follow a different course.

Homosexuals

Gays and lesbians in the military provide the simplest case: they continue to be a quasi non-issue. No figures or estimates are available as to the proportion concerned. Disciplinary incidents involving homosexuals have not attracted attention in the past few years, and there is reason to believe that the situation is essentially unchanged as regards cultural orientations, institutional norms and behaviour. Homosexuality remains firmly in the private sphere, and silence or discretion continues to be the rule as long as it does not interfere with the expected discharge of duty. Only personal attitudes and changing perceptions of its public acceptability seem to have evolved in a more liberal direction.

One interesting development, though, was the emergence in December 2001 – the very moment when the last conscripts left the services – of a French Association of Gay and Lesbian Service Members (AFMGL), whose declared objective was to fight for equal rights as well as against homophobia and discrimination through legal pressure and media publicity. However, its membership, according to data volunteered by gay and lesbian websites, never rose above 100, and it does not appear to have been active (at least on the Internet) beyond late 2003. Another such association (AD2– MIGALE) was created in December 2005, with a similar outlook. Its forum site boasts 500 visits after two months of existence – at most 0.15 per cent of total active manpower levels, with no guarantee that the figure does not include veterans, outside sympathizers or even idle visitors. This strongly suggests that discrimination against gays and lesbians is too low in the services to elicit a much stronger organized response from its actual or potential victims, and/or that their conservative orientations predispose them to put up with existing military norms and the traditional lack of symmetry in lifestyles between heterosexual majority and homosexual minority.

Another comment is in order. The two associations claim that they are modelled upon similar organizations in the United States, Canada, Britain and Germany – countries with cultural traditions far removed from France's. There is indeed an element of ambiguity in the ends pursued, which may find its origin in those models. While the declared objective seems firmly grounded in a 'modern' logic of integration (insistence on rights, freedom from discrimination, and institutional blindness to sexual preference), other aspects of both associations' statements of goals are directly or indirectly related to a post-modern logic, as for instance when themes like recognition, homosexuality as an open lifestyle, mutual assistance of gays and lesbians in uniform, and the right to communicate freely through bulletin boards are broached. The same goes for websites carrying pictures of explicit same-sex scenes among service members in battle fatigues and combat boots. The emergence of an identity built around the expressive meaning attached to minority sexual preference comes clearly through the contents of those websites. The claim for public recognition is clearly at odds with the implicit 'don't ask, don't tell' principle at the heart of dominant practice.

Homosexuality in the armed forces is too marginal to disturb them seriously, either in terms of effectiveness (issues like the disruption of primary group cohesion or the problems, real or hypothetical, raised by living in close confines are nowhere to be found in the few public comments we have come across) or in terms of external legitimacy. Yet it has the potential to stimulate cultural change in the military by introducing touches of post-modern tendencies. The statement in 2000 by the general in charge of the Army's public relations agency that gays and lesbians are welcome anywhere in that service seems a case in point.

Servicewomen

The salient fact here, in keeping with predictions, has been the dramatic rise, during the transition period and the first few years of the all-volunteer force, in the numbers and proportion of female service members. Testifying to the difficulty in recruiting (especially qualified) male volunteers to the rank and file under that format, the number of females has increased by more than half to 33,000 (44,000 if the Gendarmerie is included). From less than 5 per cent overall a decade ago, owing to the presence of over 50 per cent of conscripts who, by legal definition, were all male, the percentage of women in uniform has increased to 13 per cent. Nor is this the final word: for the past few years, the rate of female recruitment has reached 20–23 per cent, which means that the latter figure (or a slightly lower one, as women tend to serve for fewer years than do men) will soon apply to the overall composition of the armed forces.

Differences among services are significant, with the Air Force (18 per cent) leading the pack, far ahead of the Army, Navy and Gendarmerie (11 per cent each). Fully 47 per cent of medical service doctors are females. They also represent 17 per cent of privates and corporals or equivalents, 10 per cent of non-commissioned and junior officers, but only 4 per cent of senior officers. Yet the percentage of female regular officers is slated to rise strongly over time as recruitment to service academies (one in six at Saint-Cyr in the latest intake) has suddenly surged. While the proportion of female volunteers for combat positions or the combat arms remains marginal, one in six servicewomen (one in five in the Army) has taken part in the type of overseas operations that has become so common in the post-Cold War era, where the probability of inadvertent exposure to the risks of military action is far from nil.

A high proportion of women are pre-socialized to a military environment, as some 25 per cent come from military families, and close to half have relatives, husbands or partners in the forces. Hence a continuation of the pattern described and analysed earlier, mostly marked by conservative orientations, exemplified by career motives which give pride of place to the military's image as one of integrity, discipline and obedience, equity, hierarchy, patriotism, public service mission, and prestige. The armed services are seen as a perennial institution, whose function, highly structured organization and meritocratic principles guarantee social standing and secure employment. However, it is also perceived as a challenge, and a means to avoid the types of humdrum jobs women with only average education seem destined for in civilian life. Extroversion and a taste for change, travel and outdoor activities are mentioned in many an interview.

Working in a predominantly male environment is also attractive to many females, and is frequently reported as one of the reasons for joining up. Military women very often seem to enjoy their rather exceptional condition as a somewhat pampered minority, one that is valued as a source of quality personnel in an institution that (contrary to outside expectations) succumbs to misogyny less than other occupations, and scrupulously sticks to the principle of equity. One revealing element in that regard is the mixed attitudes expressed towards the prospect of an increase in the number of women: while they expect such an increase to lessen some of the difficulties they encounter, existing servicewomen also tend to fear it, on the grounds that more females would mean a lowering of standards and a deterioration of their positive image.

That is not to say that all is well: servicewomen do have their complaints. The main source of problems mentioned in interviews relates to the degree of acceptance of uniformed women by their male colleagues, among whom virility tops the value system and is part of cherished traditions (especially at service academies and in the combat arms, where the very presence of women in the group is perceived as a frontal attack on the majority's identity). Then comes the stereotyping of women's military and physical abilities as intrinsically inferior. While this is not a widespread attitude, it does mean that women have to prove themselves in core activities. The practice of sparing women certain tasks (e.g. carrying heavy loads, guard duty) is seen by some as proof positive that they make the burdens of service heavier for their male counterparts. The same applies to remarks on the averred fact that, owing to pregnancies, a higher sickness rate, etc., it takes 1.15 women to accomplish one man's normal workload.

Leadership responsibilities may raise specific difficulties for servicewomen in environments where virile values are at a premium, though the distribution of assignments and occupational specialities suggests this is not highly frequent. Given the average rank and the types of activity far removed from combat they normally engage in, the leadership they exercise is closer to that which prevails in civilian organizations. Their recognized occupational abilities, which derive from both their dedication and their higher educational attainment, more than compensate for vulnerabilities whose scope has been decreasing over time.

Marital status shows huge differences between men and women in uniform: among the latter, 53 per cent are bachelors (versus 38 per cent) and 7 per cent divorcees (versus 3.5 per cent), attesting to a particular difficulty in reconciling service and family life among females.[11] Maternity is no ground for dismissal, and legislative measures protect the right of women to six weeks' leave before and after childbirth. But it is apt to disrupt work schedules, and male superiors have been known to apply informal pressure on women to postpone it. This, as mentioned earlier, is the most popular solution among servicewomen, as indeed with gainfully employed Frenchwomen in general. Motherhood is also, in more symbolic terms, what reminds women and men of their basic difference, and leads some to wonder whether it is appropriate for those who give life to assume killing roles, thereby further questioning their true military status.

A feature that is conspicuously absent from interview material is the issue of sexual harassment. Statistics of disciplinary incidents bear few traces of complaints from women over that issue. The same is true of overt sexism, whose institutional variety has never been particularly strong and has virtually disappeared thanks to the pragmatic ethos that pervades the official doctrine of today's armed services. Its individual, deviant variety is still there, but it is probably significant that it no longer feels free to express itself openly. In any case, servicewomen do not denounce it as pervasive or particularly offensive.

In the face of such enduring difficulties, women tend to react in one of two ways to the slight margin of tolerance they sometimes 'enjoy' with regard to military codes, as when it comes to matters of dressing, hairstyles or make-up. Some deliberately play down their feminine identity and ask for equality of treatment with men; others use what little opportunity there is to emphasize it. These diverging attitudes reveal a basic tension between identity and equality typical of minority statuses.

Mutual comparison of men and women in the armed forces classically produces two types of discourse. The first emphasizes individual variability of abilities and attitudes, and therefore plays down gender differences. In that view, merit and individual inclinations should govern the assignment of positions in terms of both roles and rank, which would leave the field open to women who aspire to slots hitherto reserved for men, and implicitly affirms the illegitimacy of the few remaining bars to their self-actualization. This is despite the realization by many that full gender equality is unlikely in the armed forces.

The second, on the contrary, puts the stress on differences in aptitudes between the sexes, both physical and psychological. Women are recognized as being mentally agile, precise, dedicated, correct, tactful and persistent, but also as physiologically vulnerable, less strong physically, less consistently available for duty, and apt to crack up when things come to a head. While this line of argument avowedly proceeds in terms of stereotypes, it does have a semblance of statistical relevance, which is used to uphold, or to accept, the principle of a gender division of labour that would assign males to operational activities and females to support tasks, broadly defined. Already, in interview material, the idea pops up of a segregated division of labour which would assign administrative, technical or specialist tasks to females, and reserve 'tough' operational jobs for males – however unrealistic this may be in view of the nature of military action today, of the shares of personnel involved in each category, and of the basic organizational need for integration.[12] It would be wrong to assume that the proponents of this second view are mainly conservative men: some females say they would favour rewriting service rules in gender-specific terms, so as to introduce more flexibility in service schedules and take into account the larger share of family roles they assume.

Some women are of the opinion that the gender division of labour debate has been made irrelevant by what Edward Luttwak has called 'post-heroic warfare'[13] and by peace support operations that no longer involve the same degree of direct, violent contact with an enemy at individual level, and provide more opportunities for women to assert their abilities, distinct or not. They emphasize that servicewomen now undergo the same basic and continuing training as males. Whereas women do not flock *en masse* to operational jobs, they do resent the idea that they may not be allowed to select that option if they so wish. The amount of resentment is low, however, if only because the gradual opening of once all-male service components has accelerated in the past few years.

One argument that fails to appear in interviews is the impact of women on group cohesion and male bonding. This absence betrays lack of actual experience of the problem where it really counts – that is, where the requirement for all-round cohesiveness (not just task-oriented cohesion) exceeds the needs experienced in civilian organizations. The fact that few women serve in combat units is enough to account for such a state of affairs. Another factor is that there is no consensus among women on where, as between equality and identity, to place the thrust of further emancipation – which they desire in principle but far from always for themselves.

Ethnic minorities

Predictions had it right: the proportion of members of ethnic or racial minorities has risen – for the very same reason as that behind the increase in the numbers of servicewomen. Although such an assertion is based on estimates that are imprecise at best, the figures bandied about are somewhere between 10 and 20 per cent, comprising mostly North African Arabs plus blacks (and a few Asians) of more recent immigration waves. Interestingly, this state of affairs replicates the experience of the few all-volunteer Army battalions that existed under the mixed format. Given that the estimated proportion of French citizens of immigrant origins used in most studies amounts to 7.4 per cent of the general population, it is difficult to determine the extent to which they are over-represented in the armed forces (especially as the baseline figure itself may be grossly lower than is really the case). But with unofficial discriminatory practice now under press and government fire, there appears to be ample room for growth in their numbers.

Distribution across ranks shows them to be concentrated at the bottom: their presence is the services is too novel for them to have risen through the hierarchy. Another factor resides in the fact that many are school drop-outs for whom service under arms is seen as a second chance. The Navy and Air Force actually introduced programmes, complete with advertising campaigns, designed for low-qualified young males, which specially target them. They have learned to de-emphasize that policy, however, as its results, marked by disciplinary incidents, low effectiveness and premature separations, have fallen short of expectations.

Many regard the military as a potential avenue of upward mobility (which may explain why programmes with a strong welfare slant aimed at them do not work as well as hoped), and a source of immediate financial independence. The massive unemployment that affects them is certainly to be counted as one of the major push factors. Their other reasons for joining (prestige, respect, structured environment, etc.) are to a large extent similar to those observed among servicewomen. Patriotism is often mentioned in interviews, as is the wish to escape the kind of pervasive discrimination against them to be found in the private sector. Not unlike the example of African Americans who joined the US military, they are after 'first-class citizen' status. This places them firmly in a 'modern' logic of integration.

Yet while they wish the military to be blind to their differences, the mirror image they experience is for them a major source of frustration. Their military environment tends to over-emphasize their Islamic roots, which – in a context where Al-Qaeda terrorism abroad, burning cars and worsening riots at home figure prominently – often means that their allegiance to the country is questioned (unduly, as it happens, because potentially divided loyalties are restricted to the hypothetical event of French forces being engaged against Muslim forces in their countries of origin, in which case they would rather not be involved in the fighting). Despite the fact that there is little religious fervour among them, they find the absence of imams in uniform and of halal menus or rations galling. And because they mostly tend to shun alcoholic beverages, they feel systematically excluded from the drinking parties in which cohesiveness often expresses itself.

Suspicion on the part of others and frustration on their part are apt to lead to feelings of estrangement that they resent. Furthermore, like female service members, they constantly have to prove themselves in core activities, and be on the tip of their toes most of the time. Stereotyping, racist jokes or barbs, harassment, discrimination (informal rather than formal) are not unheard-of, and little effort is made to put their skills (notably their command of Arabic) to good use.

Some find it hard to bear, and leave prematurely. Most stay and soldier on, however, without resorting to the kind of defensive communal feeling one might expect when a sizeable minority is made to feel different in spite of its declared wish to be part of the larger group. In other words, they – for the time being – reject the post-modern logic to which their plight may one day conduce.

In the face of a situation now on the verge of becoming critical, the system's leaders seem ready to concede that something has to be done to accommodate the needs and aspirations of a group that is here to stay. Indeed, as of this writing a decision has at long last been made to create a Muslim chaplaincy. The urban riots of autumn 2005 which prompted the government to declare a state of emergency surely helped dramatize the need for change after years of cultural inertia and complacency. As a result, things are definitely on the move – in both the military and society.

Conclusions

Owing to its vulnerability to shortages of accessions by mainstream young males, the all-volunteer force has soon learned the value of attracting recruits from pools it had hitherto not tapped to any great extent. Together, women and the children or grandchildren of immigrants are now well on their way to representing fully 30–35 per cent of the force in the not too distant future. Homosexuals cannot be considered a group to be tapped in any formal sense, and while they do not appear to be numerous, or willing to claim full recognition, the lifting of social inhibitions in the parent society (as well as the voluntary logic of an all-volunteer force) renders it likely that they will make themselves heard, however softly. Despite the tendency of French culture to live with problems and hope for the best rather than effectively solve them, the needs and aspirations of all three groups will have to be taken care of. The process of accommodation has already started.

There are interesting contrasts in the manner in which that process is unfolding among the three categories under consideration. Where women are concerned, it started two decades ago and has been characterized since by patience and the apparent wish to avoid antagonizing the dominant military culture. It is well advanced, and all that remains is for them to reap the benefits to which their fast-rising numbers will soon entitle them. Minorities, especially the Muslim component among them, seem set to follow the same course – with one major difference: their plight has recently attracted political and media attention. The recent urban riots have put the spotlight on the amount of discrimination they suffer in society, and the armed services now appear as an institution that can help them erase the stigma of second-class citizenship. Outside support guarantees that the process as far as they are concerned will in all likelihood gain momentum faster than has been the case with women. Homosexuals play too marginal a part to exert a decisive influence. But all three cases, one element is in common: the restraint with which aspirations for change are expressed. Changing a military culture is a slow business.

Quite obviously, post-modern trends have so far failed to penetrate deeply into the military fabric: the cultural tradition of de-emphasis of internal differences accounts for that. Separate platoons or equivalents for women or descendants of immigrants are unlikely in the French context. A safe bet is that the kind of know-how developed by the Foreign Legion for turning a highly diverse force into a cohesive whole will be applied instead. But that is not to say that the post-modern logic is completely absent from the new scene.

For the sake of analysis, the two logics have been described here as antithetical. Yet, in real life their relation need not be an either/or proposition. They can be combined, especially when one is pursued as an end, the other as means to that end. That, precisely, seems to be the case today in the French military. Traces of post-modern attitudes do exist. Some servicewomen would welcome rewriting service rules in gender-specific terms, and favour female identity rather than equality with men. Muslims expect the military to show a minimum of respect for the most basic features of their identity (no pork in menus or rations, military imams, etc.). The very few gay and lesbian militants want recognition of open homosexual lifestyles as legitimate. But such tendencies do not reflect majority feeling, and are mainly aimed at ending the discrimination they suffer in the existing state of affairs.

This is leading, on the basis of functional arguments, to a redefinition of what is acceptable and what remains culturally out of bounds. If the most recent trends are any guide to the future, some differences separating women, minorities and homosexuals from the mainstream will be absorbed. This partial bridging of differences is what managing diversity is all about. The most radical forms of post-modernism will remain alien to the military.

The pace of change has accelerated since the advent of the AVF – the key hypothesis of the new format as its main driver thus finds itself validated – and further change is on the cards. This in turn raises new questions for military personnel managers. What will be the effect of such change on the mainstream pool of enlistees? Will over-representation of racial minorities deter old-stock young males from joining? Will accommodating the needs and aspirations of women, minorities and homosexuals put off those traditionally attracted to the military precisely because it is conservative and unlike society? If so, will it make the risk of recruitment shortages worse? Such issues strongly suggest that freedom of action may be limited for the system's leaders in future. But, as Kipling would have said, that is another story.

Notes

1 M. Ozouf, *Les Mots des femmes*, Paris: Fayard, 1995. (See especially the concluding chapter: 'Essai sur la singularité française'.)
2 It should indeed be noted that race is not the sole issue here: there is much overlapping of Third World origins, more or less visible racial traits, and Islamic culture. The latter is all the more regarded as an alien phenomenon because France has grown to be a highly secularized society in which religious fervour influencing all or many aspects of life is mostly to be found among Muslims.
3 Ph. d'Iribarne, *La Logique de l'honneur*, Paris: Seuil, 1989.
4 M. Crozier, *Le Phénomène bureaucratique*, Paris: Seuil, 1963.
5 See Charles Moskos's analysis of the American case in Chapter 2 of this book.
6 G. Robert and L.M. Fabre, 'Homosexualité et armée', unpublished document, Paris: CRH (Army Staff), 1993.
7 However, the status of over-representation in the new equation remains uncertain. Too close an association of the armed forces with any one particular identity to the point of crowding others out may carry politically sensitive consequences, especially if casualties are suffered or even expected in an operation. Discouraging other groups from enlisting through image distortion may result in recruitment quality problems. Also, the dominance of a given identity carries the risk of introducing biases in the way missions are interpreted and performed, hence making for armed forces that are less functionally effective and politically responsive.
8 Separating recruits by gender at the most basic level of training and housing them in separate barracks was recommended in the United States a decade ago by a special panel headed by former senator Nancy Kassebaum Baker (Rep., Texas) in drafts of reports prepared for the Department of Defense. See D. Priest, 'Panel Urges Segregating Military by Sex', *International Herald Tribune*, 17 December 1997.

9 The classics of military social science, notably those of World War II fame (Grinker and Spiegel, Marshall, Shils and Janowitz, Stouffer *et al.*) did not seem to harbour any doubts (a) that primary cohesion is the mainspring of combat (or, more generally, military) effectiveness; and (b) that homogeneity enhances it, while it is diminished by diversity (of age, religion, regional attachments, etc.). Two or three decades later, authors such as Baynes, Moskos, Richardson, Wesbrook, Sarkesian, Kellett, Henderson, Segal and Kinzer-Stewart came to the same conclusions.
10 The treatment to follow relies heavily on C. Withol de Wenden and Chr. Bertossi, *Les Militaires Issus de l'Immigration,* Paris: Institut françis des relations internationales/Centre d'études en sciences sociales de la Défense, 2005; B. Boëne, 'The Changing Place of Servicewomen under France's New All-Volunteer Force', in G. Harries-Jenkins (ed.), *The Extended Role of Women in the Armed Forces,* London: European Research Office of the US Army, and Centre for Research on Military Institutions, University of Hull, 2002, pp. 8–95; Observatoire de la féminisation, *Regards sur les femmes militaires,* Paris: Ministèe de la Défense (DFP-SGA), 2005; Observatoire social de la Défense, *Bilan social 2004,* Paris: Ministèe de la Défense (DFP-SGA), 2005; Saïd Haddad, 'La Culture militaire àl'épreuve de la professionnalisation: quelques pistes de réflexion', in Françis Gresle, *Sociologie du militaire,* Paris: L'Harmattan, 2005.
11 This can be better understood if one bears in mind that the gender division of labour in French households, as national data make abundantly clear, is still very unequal, with wives assuming responsibility for most domestic chores and childcare (which may explain why day-care centres are uppermost in servicewomen's minds).
12 This idea of a 'two-tier military', in which the support functions would be allowed to converge further with civilian organizations while the combat components would continue to diverge from them, goes beyond the issue of women's roles. Such an idea brings to mind the concept of a 'plural military' which emerged in the United States precisely at the time when its armed services did away with conscription: see C. Moskos, 'The Emergent Military: Civil, Traditional, or Plural?', *Pacific Sociological Review,* 16, 2, 1973, 255–280; W. Hauser, *America's Army in Crisis,* Baltimore: Johns Hopkins University Press, 1973; Z. Bradford and F. Brown, *The United States Army in Transition,* Beverly Hills, CA: Sage, 1973.
13 E.N. Luttwak, 'Towards Post-heroic Warfare', *Foreign Affairs,* 74, 3, 1995, 109–122.

12 Diversity in the German armed forces

Heiko Biehl, Paul Klein and Gerhard Kümmel

Introduction

When searching for information about minorities in the Bundeswehr, one will be puzzled by the fact that, generally, information is available concerning women in the armed forces, but not concerning other minorities. The lack of research on ethnic/cultural and national minorities is first of all due to the fact that, for legal reasons, all the soldiers in the Bundeswehr are German citizens. Moreover, it is rather difficult for foreigners to obtain German citizenship. Finally, in the Bundeswehr people who belong to ethnic or cultural minorities usually function in a rather isolated way and not as groups. That makes it even easier to ignore the few problems they might create. The rare attention paid to the issue of homosexuals in the armed forces, by contrast, is mainly due to the fact that for a long time their presence has been regarded as a taboo, thus making it almost impossible for scientists to gain access to the respective data.

With women, who formally gained access to the military in the mid-1970s and who were initially only permitted to join the medical service and military musical bands, the situation is quite different. Here, there was some public as well as scholarly interest and debate which resulted in a number of publications on the topic. There was a futher burst of interest when women were granted full access to the armed forces on a voluntary basis in 2000.

In this chapter we will deal with these three minority groups in the Bundeswehr in the order that they have been mentioned in this introduction. On the basis of our findings, we will end our analyses with a brief account of what to expect in the future. We will start however, with a brief look at the national and ethnic/cultural composition of German society in order to contextualize our discussion of these minority groups in the Bundeswehr. Soldiers from the former East Germany, who are often perceived to be a minority in the Bundeswehr, are not our subject here. They have become so numerous in the armed forces that it is no longer justified to consider them a minority.[1]

National and ethnic/cultural minorities

With regard to ethnic, cultural or national minorities in Germany, one generally distinguishes between four groups, which can again be subdivided. The first and smallest group is made up of minorities that are officially recognized by the Federal Republic of Germany. Foreigners who live in Germany but do not possess German citizenship constitute the second group. A third group is made up of naturalized foreigners. The fourth and last group comprises resettlers or late resettlers of German descent.

The officially recognized minorities in Germany are approximately 30,000 Danes in Schleswig-Holstein, about 60,000 Sorbs in Brandenburg and Saxony, and roughly 30,000 Sinti and Roma who have settled throughout the entire territory of the Federal Republic. They are German citizens and have the same obligations and rights as any other German, but enjoy some special rights with regard to their cultural life and customs in particular, but also to their language.

Among the approximately eighty-two million people who nowadays live in the Federal Republic of Germany, about 7.3 million are foreigners. Though they come from almost all countries in the world, most of them are of Turkish descent (1.9 million), followed by people from former Yugoslavia (0.6 million), Italy (0.6 million), Greece (0.3 million) and Poland (0.3 million). Furthermore, Spanish, Portuguese, Russian, Vietnamese, Moroccan, Persian and Lebanese people are to be found in considerable numbers. Foreigners are neither enfranchised nor eligible to be; not being allowed to become public officials or professional soldiers, they also are not subject to compulsory service and thus not to be found in the Bundeswehr.

The legal situation up to 2000 meant that the naturalization of foreigners in Germany was linked to conditions that were difficult to comply with. German citizenship was not to be obtained in a quasi-automatic way but only upon request. The applicant had to have lived in Germany for a long period, and had to show sufficient command of German, as well as acquaintance with German culture. So, the number of foreigners who obtained German citizenship was rather low. Approximately 253,000 people were naturalized in Germany from 1986 to 1994. In the same period 537,000 people in Great Britain were granted citizenship, 486,000 in France, 247,000 in the Netherlands and 228,000 in Sweden.

Because of this policy, from 1972 to 1999 only 750,000 immigrants were granted German citizenship, thus becoming either Germans without German ancestors or Germans of non-German descent.[2] Since 1 January 2000, naturalization has been facilitated, though, owing to some reforms of the German citizenship code. The number of years passed on German territory required in order to be entitled to claim German citizenship has been reduced from fifteen to eight. Children born in Germany whose foreign parents have been living in Germany for at least eight years obtain both German and their parents' citizenship on the day of their birth and have to decide on one of them when coming of age. With the introduction of these laws, more than 500,000 foreigners in Germany have been naturalized up to the middle of 2003. This number is more than twice as high as the number in the corresponding preceding period.[3]

Resettlers or 'late' resettlers[4] of German descent *de jure* do not constitute national minorities. They are the descendants of Germans who emigrated from the Middle Ages onwards to countries in East and South-Eastern Europe. There they lived as minorities among the local population and, to differing degrees, adapted the customs of their new environment while preserving their old German peculiarities. After the Second World War most of them began to return to Germany because of adverse living conditions or because they had become unwelcome in these countries. In this, they were supported by the German government.

In the Federal Republic of Germany these resettlers are considered to be Germans and are entitled to claim some automatic fast-track naturalization. However, they often do not speak German very well, and may have preserved the habits and customs of the country from which they emigrated, as is the case with any foreigner. Similarly, to their fellow citizens without German nationality they sometimes represent 'alien' elements among the old-established German population and are regarded as 'Russians', 'Poles' or 'Romanians'.

The integration of resettlers into the society of the Federal Republic caused little trouble until the mid-1980s, since the annual number of immigrations was low, often being less than 50,000. This situation changed in 1988, however, when for the first time a quota of more than 200,000 resettlers was reached. From then on, integration proceeded less smoothly, and the resettlers had to cope with certain difficulties. Not only were they regarded as annoying competitors for the gradually diminishing number of jobs, but they were also subsumed into the category of 'foreigners' by a majority of the German population. In an inquiry carried out by the Bundeswehr Institute of Social Research (SOWI) in 1999, only one-quarter of the interviewees accepted late repatriates as Germans, whereas nearly 55 per cent denied their German citizenship.

In addition, an ever-growing number of repatriates immigrated to Germany

Table 12.1 Number of resettlers immigrating to Germany, 1981–2004

Year of immigration	Number
1981–83	107,867
1984–86	118,275
1987–89	658,251
1990–92	850,032
1993–95	647,749
1996–98	394,928
1999–01	267,878
2002–04	196,301

Source: Federal Administration Authority (Bundesverwaltungsamt), Cologne.

with an ever-diminishing knowledge of the German language. As a result, in 1996 the federal government felt obliged to introduce a linguistic test as part of the proceedings to obtain German citizenship. However, only late repatriates themselves have to take this test, not their children, who therefore very often grow up with a less than perfect knowledge of the German language.

National and ethnic–cultural minorities in the armed forces

As a result of what has just been sketched, a soldier on active duty in the German armed forces who bears a foreign name, has a different skin colour or belongs to a non-Christian, usually Muslim, religious community is either a member of a sustained minority of German nationality, a naturalized foreigner or a resettler of (former) German nationality/descent.

The members of sustained minorities have the same rights and duties as other German citizens. Thus, they may become soldiers and are subject to compulsory national service. As they speak German fluently (apart from their native language or idiom) and do not cultivate excessively deviant manners and customs, they do not attract attention in the ranks. Sinti and Romany are scarcely to be found in the Bundeswehr, since, generally speaking, they are not very much interested in a volunteer service – and as for compulsory service, they fall under the exceptional regulation that descendants of victims of National Socialist persecution may be exempted from military service.

Some of the former foreigners have applied successfully for German citizenship and some Jews of German nationality in the meantime do their military service or serve as volunteer soldiers in the German armed forces these days. Because of their negligible absolute number and the fact that, as a rule, they have completely adapted themselves to German manners and customs, their presence in the armed forces is mostly not accompanied by severe problems. There are just occasional exceptions regarding Jews and Muslims. As it is evidenced by some petitions addressed to superiors and also to the Defence Commissioner of the Bundestag, there are sometimes difficulties arising from observance of alimentary rules, from the peaceful exercise of religious rites and from holiday regulations. This is partially due to the fact that at ground level regulations already in foce guaranteeing observance of dietary rules for each soldier in compliance with his religion are often ignored.

There are still no regulations concerning different holidays. Also, there are no central regulations regarding the installation of devotional rooms for each religion, or rules regarding particular religious service times and horary prayers. All this is the responsibility of the garrison commander. Nevertheless, a superior can only comply with the religious needs of his subordinates when he has been informed of their religious affiliation. So, the subordinates have to become active, since a soldier's roster shows only his affiliation to one of the important Christian religions, but not to other religions or religious groups such as Muslims or Buddhists.

In 2000 the coalition government in Germany agreed upon making the naturalization of foreigners easier, and even introduced provisions for dual citizenship. Thus, in time the Bundeswehr will also encompass a greater ethnic, cultural and religious variety. In particular, the number of Muslim soldiers will increase.

Jews and Muslims will both insist upon observance of their religious rules. The Bundeswehr, especially as the basic right of freedom of religion is concerned, will have to comply with these demands. Currently under discussion is the question of whether or not, apart from Protestant and Catholic military chaplains, there should be chaplaincy services for other religious groups in the Bundeswehr in the future. At present, religious ministration to Muslims and Jews by their own clergy is still prevented by the fact that, according to a regulation in force, a corresponding chaplaincy service only can be claimed when the number of soldiers affiliated to the religious community concerned exceeds 1,500 in a garrison. Nevertheless, it is agreed that Bundeswehr soldiers affiliated to non-Christian religious communities may in no case be forced to celebrate Christian holidays. So far as their own religious holidays are concerned, they are dependent on arbitrary decisions of their superiors. A corresponding holiday regulation is to be expected, agreed upon in compliance with the central offices of each religious community concerned.[5]

In spite of the small number of Jewish soldiers in the German armed forces, the Jewish Central Council recently designated an officer of this religion as a contact person to meet the special needs of Jewish soldiers in the Bundeswehr. As yet there is no information on racist-rooted or anti-Semitic incidents within the Bundeswehr. Obviously, the few non-white and Jewish soldiers actually in service meet with no difficulties.

However, there are sometimes problems with regard to the resettlers. Their number in Germany runs to more than four million, and they come rather exclusively from East European countries. Because of their German ancestry, the resettlers are conferred full civil rights and duties upon arrival in Germany. They are also liable to conscription despite their often deficient knowledge of the German language. If some of them are conscripted in the same unit, they often form a distinct sub-group, separating themselves from the other soldiers. The latter used to call them 'the Russians', making them the subject of prejudice, exclusion and suspicion.

Because of their poor German, sons of resettler families are often allotted simple functions in the Bundeswehr. Thus, certain infantry units of the army and ground units of the Air Force sometimes receive a disproportionally high number of resettler conscripts. In some cases this fact has already led to difficulties in superior–subordinate relations, but also within the rank and file.

A SOWI inquiry among unit leaders of the three services in 2000 dealt with the problems of minorities in the German armed forces.[6] The report based on this inquiry confirmed, from the perspective of 209 officers commanding a battalion or similar unit, the results mentioned earlier with regard to cultural and religious minorities, but also underlined the problems linked to resettlers and late resettlers. They were present in three-quarters of the units whose commanders had been interviewed. Members of the Jewish faith, on the contrary, were only to be found in every twenty-fifth unit, whereas followers of other non-Christian religions were only present in every third unit.

The unit leaders commanded 112,698 soldiers altogether: 2.8 per cent of them belonged to the group of resettlers, which had a particularly high representation in the army. In about one-fifth of the units with repatriates they made up 5 per cent of the soldiers. In approximately 6 per cent of those units, their number exceeded 10 per cent. In one case their proportion reached as high as 16 per cent. In terms of quantity, the repatriates thus constituted the most significant group of the minorities inquired. Their presence on duty was an everyday experience for most of the units. The difficulties linked to their presence, however, reached an extent not to be found with any other minority.

According to the findings of Kümmel and Biehl, [7] three main areas of problems were to be distinguished with regard to the group of repatriates: their lack of language skills often makes their integration into the military order more difficult, they are often socially isolated from their fellow soldiers and they are more often liable to military misbehaviour than the other soldiers. Thus, 30 per cent of the unit leaders report a strong tendency towards heavy drinking for subordinate late repatriates. In addition, the interviewees said that a disproportionately high number of late repatriates violated the federal drug regulations. One of the interviewees even cited cases of illegal procurement of goods, which was partially accompanied by the use of violence. All in all, 45 per cent of the unit leaders interviewed reported a tendency for late repatriates to use violence in the settlement of conflicts.

With regard to the second great problem linked to late repatriates in the armed forces, it was evident that large parts of their socialization had taken place outside Germany. Almost 60 per cent of the unit leaders commanding late repatriates reported an insufficient command of the German language among those soldiers. This caused linguistic barriers, which had some effect on the daily service routine. In one-quarter of the units, the formation and command were adversely affected. The language was not the only problem for the late repatriates' integration into the armed forces, though. In 17 per cent of the units it was reported that difficulties in the military formation and command were also due to a lack of educational and professional skills.

It is not only relations with their superiors that seem to be affected by this barrier, though. More than one-third of the soldiers interviewed lamented the late repatriates' lack of integration with their comrades. This was due to a combination of social self-isolation and avoidance by their comrades. Some 13 per cent of the unit leaders even stated that it was very hard to integrate the late repatriates into the military order.

The difficulties linked to large concentrations of late repatriates were especially serious, according to the unit leaders surveyed. In their opinion these concentrations facilitated the development of cliques. They even cited rare cases of gang formation and the development of a kind of Russian mafia. Statistics showed the greatest difficulties to occur in units marked by a higher proportion of late repatriates. This tendency was to be perceived in all the areas mentioned: a higher percentage of late repatriates in a unit increased the risk of problems with the maintenance of military discipline and caused difficulties with regard to integration into the daily service routine, resulting in the social isolation of the late repatriates. As was mentioned above, the latter process sometimes led to the development of groups and cliques.

Homosexuality and military service

As is the legal situation in nearly all European countries, homosexuality is not subject to punishment in Germany, either in its male or in its female form. This is also an orientation value for the Bundeswehr. Thus, homosexuals are allowed to become volunteer soldiers and they also are liable to conscription. Current accounts see a ratio of about 5 per cent of open homosexuals living in registered household partnerships in Germany. As for the corresponding predisposition or orientation, estimates reach up to about 12 per cent of the male population. Unfortunately, there are no data available about the number of homosexual soldiers in the Bundeswehr, and there is also no information about whether their number is larger or smaller than the national average. Because of the Defence Ministry's refusal of an in-depth examination of the issue of homosexuality in the armed forces, no valid information at all has been published on this topic as yet.

Although homosexuality is not punishable, the personnel management of the Bundeswehr, for a long time, started from the presumption that a gay soldier would be restricted in his aptitude, and capacity for duty. The validity of this opinion was even backed by a supreme court decision of the Bundesverwaltungsgericht (Federal Administrative Court). In the past this meant in practice that most homosexuals, though liable to conscription, were not called to arms when their sexual orientation became known. Candidates volunteering to join the Bundeswehr were still refused if they confessed their homosexuality during their preliminary medical examination.

If a soldier's homosexual orientation became known when he was already on duty, investigations took place to discover whether he had had sexual relations with (an)other soldier(s). Unlike in the case of heterosexual relations, it made no difference whether these sexual contacts took place during duty or leisure time. In either case, the soldier was liable to disciplinary action. According to the data available, between 1981 and 1992 the disciplinary courts took disciplinary actions against sixty-three soldiers for homosexual behaviour.[8] The sentences passed ranged from nine discharges from service to twenty demotions to twenty-one bans on promotion. Seven acquittals were pronounced, and a fine was imposed on the rest.

Even if there were no active homosexual relations, the soldiers concerned had to count on disadvantages. The military command presumed that homosexuals generally were not fit for promotion without restrictions. The assumption was that their sexual orientation, once revealed, could lead to a loss of authority, thus negatively influencing the troops' operational readiness and discipline. Furthermore, claims were to be heard that gay soldiers, because of their sexual orientation, could easily become the victims of blackmail by other soldiers, or even by hostile secret services.

When a soldier's homosexuality was revealed, the soldier concerned, if he was in a position of superiority, usually lost his command status. Instead, he was assigned activities that could be done without any subordinates. He was excluded from any future command and instruction duty, as well as from access to secret documents. Between 1990 and 1997 six investigations took place in the Bundeswehr against soldiers solely because it became known they were gay. In four cases the men concerned were refused the chance to become professional soldiers. Two of these four soldiers were released from their command assignments; another was turned down for a long-planned new posting.

The situation has since changed. In early June 2000 the then Defence Minister, Rudolf Scharping, declared that homosexuality no longer justified restrictions with regard to a soldier's employment and status. In December 2000, General Kujat, the Chief of the Federal Armed Forces Staff at the time, published a paper advising superiors on how to deal with questions of sexuality within their ranks. In this paper he demands respect for the individual's right of sexual self-determination, and tolerance towards other non-criminal sexual orientations, including homosexual male and female soldiers.[9] Thus, a remarkable step was taken to settle the formal aspectss of this issue, a step that also led to the foundation of the Working Group of Homosexuals in the Bundeswehr (Arbeitskreis Homosexueller Angehöiger der Bundeswehr) in 2002. This does not mean, however, that the problem of homosexuality in the German armed forces has been solved entirely, since there is still a great deal of uncertainty with regard to the correct behaviour towards homosexuals. Great efforts will be required to reduce homophobic attitudes among soldiers.

Women in the Bundeswehr

In late 1973 the German government of the time, a coalition of the Social Democratic Party of Germany (SPD) and the Liberal Party (FDP), responded to increasing demands for democracy, political participation and emancipation in society by establishing a commission of inquiry on Women and Society. One result of the commission was that the Defence Minister, Georg Leber (SPD) opened the Bundeswehr to women.[10] Thus, in autumn 1975, the year the UN had declared Year of the Woman, women entered the Bundeswehr as career officers, serving as doctors and pharmacists. This participation, however, was restricted to the medical service for two reasons. First, Germany's Basic Law prohibited any armed military service for women; female soldiers were merely instructed in the use of small arms (pistols, rifles) for the purposes of self-defence and defence of their patients – which implied that women were not liable to guard duty. Second, in the early 1970s the medical service was suffering from substantial recruitment problems, as at the time there was a gap of 1,300 longer-service volunteers for officer careers. Yet, after some time, female access was extended to military musical bands because in emergency the soldiers making up the bands were transferred to the medical service.

In the early 1980s discussion about extending female participation to further classifications gained new momentum. In 1981 an independent long-term commission had analysed the shifts and trends in the demographic composition of society with an eye towards satisfying the Bundeswehr's personnel needs. One year later, the commission submitted its report, one paragraph of which recommended that allowing women to volunteer for non-combat functions be considered. However, no such political move was made during the 1980s. Nevertheless, the issue remained on the agenda of societal and political debate.

As a result, further steps were taken to increase the representation of women in the armed forces. From the beginning of 1991, all careers in the medical and military musical service were made accessible to women.[11] In 1994, Verena von Weymarn became Surgeon General, the first female general in German military history. In the late 1990s, women made up about 1.2 per cent of all German soldiers. The number of women was about sixty in military bands and about 4,350 in the medical service (approximately 400 medical officers, 700 medical officer candidates, 2,300 non-commissioned officers, 200 non-commissioned officer candidates and 100 in private ranks).[12] In addition, close to 50,000 women worked as civilian employees either in the armed forces or in the Federal Armed Forces Administration, often on billets where other armed forces employ only soldiers.

According to official and unofficial declarations, no problems whatsoever were arising in the Bundeswehr from the presence of female soldiers. A research study among female officer candidates of the medical service, carried out by the SOWI in 1993, also led to the result that 'after one year of military instruction, female soldiers on the fields of value orientation and attitudes may be considered as integrated into the Bundeswehr'.[13]

Towards the end of the millennium, women in the Bundeswehr were still confined to non-combat roles – that is, to the medical service and to military bands – whereas in other countries the integration of women had progressed, sometimes substantially. But in 1996 a 19-year-old trained electrician, Tanja Kreil, applied for voluntary service in the area of maintenance – that is, in a combat support function. Her application was declined. The Bundeswehr argued that women serving in combat functions and with weapons in their hands were forbidden by law. Tanja Kreil did not accept this decision and took her case to court.

Part of her case was that her application had been illegally rejected because the Bundeswehr resorted to her sex and used a gender-specific argument by saying that men were allowed to enter positions involving the use of arms while women were not. The argument ran counter to a February 1976 European Union (EU) directive that requires member states to follow the principle of equality of treatment in the workplace – that is, it prohibits discrimination in the workplace for reasons of gender. Therefore, the Administrative Court in Hanover asked the

European Court of Justice (ECJ) in Luxembourg for an interpretation. Eventually, in January 2000, the ECJ considered Tanja Kreil's charge to be substantiated and the Defence Ministry was tasked to increase the number of classifications and trades available for women. This means that the recent steps to open the Bundeswehr to women to a much larger extent than before do not stem from genuinely political initiatives, as one might have thought, but from a court ruling that required the political sphere to take some action.

Following the ruling of the ECJ, there was a lively and controversial debate on the topic of women's inclusion in the military, both within the armed forces and in the wider German society. One of the issues discussed was the depth or the degree of integration and thus the question whether, taking into account the text of the ECJ ruling, certain areas, classifications or branches should be denied to women or whether the Bundeswehr should be completely open. The Ministry of Defence eventually chose the latter option – already advocated in mid-February 2000 by an expert opinion from the SOWI.[14]

Starting in January 2001, women became eligible to enter all classifications and branches. Two premises guided further steps taken to include women in the armed forces. (1) Integration incorporated the principle of voluntariness. Accordingly, women would enter the military services voluntarily and would not be subject to conscription, as is still the case for men. (2) The integration was to be based not on policies of affirmative action, but on the principle of equality of treatment. Thus, there was to be a gender-free or gender-neutral assessment of those who applied for military service – that is, everyone would have to pass the criteria for the position they were applying for, irrespective of their sex.

In the summer of 2000, the first job interviews and aptitude tests were conducted with female applicants. Simultaneously, steps which most of the people involved deemed necessary were taken to revise the German Constitution. At the end of October and in early December respectively, the German parliamentary bodies, the Bundestag and the Bundesrat, voted by two-thirds majorities for the revision of Article 12a of the Basic Law. In January 2001, women began to enter the Bundeswehr in trades hitherto precluded for females, in several cohorts, usually starting service every two months. Overall, there were about 2,740 servicewomen in 2001. As of January 2006, some 12,000 women are serving in the German military, overwhelmingly in the Army and mostly as non-commissioned officers, thereby amounting to about 6.5 per cent of all German shorter- and longer-service volunteers and career soldiers. This constitutes an impressive increase from a percentage of 1.2 per cent in the late 1990s.

The history of the inclusion of women in the military in several other countries shows that the whole process is by no means smooth, but faces substantial opposition, especially when it comes to women in combat roles. This resonates with the findings to be found in the field of organization sociology, gender studies, in feminist writings and in military-sociological research. According to these studies, once images and perceptions of the roles of women in society start to change, the gender system as a whole cannot be left untouched. Instead, the

Table 12.2 Number of female soldiers in the three German services, according to career, 2004

	Navy	Air Force	Army
Rank and file	156	218	977
NCO candidates	229	293	1,069
NCOs	734	1,220	3,455
Officer candidates	256	380	919
Officers	126	195	530
Total	1,501	2,306	6,950

Source: German Department of Defence.

change in role images for women also means a change in role images for men and for the gender system as a whole.[15] In this vein, the integration of women and, in particular, the unrestricted and complete opening of the armed forces to women, including combat functions, severely affects the commonly held and cultivated image of the soldier as a male warrior.[16] What will be the male soldiers' reaction? Will there be a polarization in gender relations in the Bundeswehr, a male 'backlash'?[17] Will the female soldiers follow a path of assimilation to the dominant culture in the organization?[18]

Recent and ongoing empirical studies have tried or are trying to shed some light on these questions. A survey of *male* Bundeswehr soldiers *prior* to the actual implementation of the integration process revealed a good deal of ambivalence, although the data in general seemed to document a relatively integration-positive climate in the Bundeswehr. Anticipated positive effects were accompanied by different degrees of reservations in some segments. Such attitudes had two sources: one was the traditional image of the soldier, the military and gender roles, and the other was status inconsistency – that is, a fear of the competition presented by women soldiers.[19] To deal with these reservations and upon recommendation by Biehl and Kümmel, the German Ministry of Defence introduced so-called gender or integration training.

A survey of *female* Bundeswehr soldiers of the class of 2001 at the *beginning* of their military career *inter alia* explored the women's motivations for joining the armed forces, their first impressions of the military as an occupational field, the areas in which they foresaw or had found possible problems and also the areas in which they felt a certain discomfort and apprehension. The findings indicated a strong inclination and motivation on the part of the women to succeed in the military. Further, they show that the formal side of integration was quite different from its social side. Indeed, the social integration of women was far from being accomplished, as in semi-structured interviews women sometimes report subtle mechanisms of exclusion used by male soldiers. For example, soldiers are the target of derogatory talk in the sense that they are suspected by men of having widespread sexual contacts.

This gained further importance in the view of obviously increasing concerns of male soldiers as to the perceived detrimental effects of female integration on military effectiveness. In addition, particular problems were expected by the servicewomen with regard to the compatibility of family/partnership and profession, requiring institutional readjustments by the armed forces (gender mainstreaming). Therefore, the study's advice was to pay increasing attention to the social side of integration as well as to somehow accommodate both 'greedy institutions' – the family and the military. In this regard, the plea was to consider steps so far unthought of, namely the introduction of part-time working schemes and childcare policies and the creation of a surplus pool of personnel that would compensate for maternity and family leave.[20]

Meanwhile, the Law on Equal Opportunity for Soldiers of January 2005 has introduced substantial elements of gender mainstreaming into the Bundeswehr. According to this law, soldiers, both male and female, are allowed to work part-time, to increase the compatibility of family and military life. In addition, the election of full-time equal opportunity representatives is currently under way. If the capabilities of a male and a female candidate for a given position are the same, the law provides that the woman should be given the job. In this way the Defence Ministry wants to increase the percentage of women in the Bundeswehr to 15 per cent in the long run.

A look into the future

As regards ethnic, cultural and national minorities, the Bundeswehr is far from being the ideal of a conscript army, namely the mirror image of society. Amid a multicultural society, the German armed forces form something like a national island.[21] This could change soon, with the full application of the nationality code adopted in 2000. As a consequence, many children of foreign descent, but born on German territory, will opt for German citizenship when coming of age and will thus be liable to military service, or will be able to serve in the armed forces as volunteers.

It is estimated that about half of the 7.3 million foreigners living in the Federal Republic these days will either apply for German citizenship or will opt for it while growing up. Among them there will be quite a few who are not part of any Christian community, but believe in other religions. This affects the Bundeswehr in so far as up to 10,000 young religious Muslims will be available every year to do their military service or volunteer for service. If they are fit for military service and do not refuse to do it, they will be drafted. Otherwise, the principle of equal treatment would be contravened. Probably few linguistic problems would arise concerning these former foreigners since most of them have grown up in Germany. Often they speak German even better than their real mother tongue. But the armed forces would have to take into consideration that most of these conscripts have other manners and customs, first of all with regard to religious beliefs and prescriptions. Certainly, many Muslims have become secularized, but a considerable minority will insist on their hereditary religious rights.

The Bundeswehr will have to comply with these facts. But here, any accommodation in the greater ethnic, cultural and religious variety purely by means of service instructions and legal regulations would turn out to be insufficient. First of all, intuition, understanding and tolerance from all the soldiers, particularly from superiors, will be much in demand.

Also, there will be calls for sensitivity when dealing with resettlers. Furthermore, language courses will be needed in the armed forces for these soldiers. The problem of resettlers within the armed forces, however, will solve itself, since their number is falling. While 397,000 resettlers were counted in 1990, by 2004 there were only 56,000.

In time there could also be a change in attitudes towards homosexuals in the armed forces. On the one hand, pressure by public opinion will contribute to this change, since homosexuality is increasingly perceived as an acceptable form of sexual behaviour. On the other hand, the legal settlements within the armed forces and the young soldiers' lack of prejudice will create more tolerance.

As regards women in the armed forces, one may state that, prima facie, their integration is proceeding quite smoothly. There are shifts in the gender system in German society. These have an impact on the armed forces, because the Bundeswehr cannot distance itself from developments going on in the wider society without paying dearly for doing so. Nevertheless, it would be too far-fetched to hypothesize or assume civil–military congruence over the long term.

A closer inspection reveals certain caveats in terms of female soldiers' social integration, which points to the need for a sustained effort to manage gender relations in the military. There is some unrest, fear and opposition among a considerable number of male soldiers in the Bundeswehr. These reservations can indeed make a difference in the further development of integration. They may translate into actual behaviour when male soldiers meet female soldiers in their workplace. Therefore, these opinions, attitudes, expectations and reservations based on traditionalism and/or on status inconsistency will have to be accommodated and managed by the German armed forces and their political and military leadership.

Part of this will be the further examination of gender images in the military, among male and female soldiers, and the analysis of soldierly interaction. But of crucial importance will be the unmistakable signalling by the political and military leadership that integration is wanted, not least because the recruitment of women alleviates the military's personnel problems. This in turn will require the organization to pay more attention to the issues of family and childcare, and flexible working hours. In the end, the successful integration of women will be contingent on the soldiers' feeling, both male and female, that the Bundeswehr is dealing with their concerns in a sincere and authentic way.

Notes

1 There is a substantial literature on this subject. See, for example, P. Klein and R. Zimmermann (eds), *Beispielhaft? Eine Zwischenbilanz zur Eingliederung der Nationalen Volksarmee in die Bundeswehr*, Baden-Baden: Nomos, 1993; S. Collmer, P. Klein, E. Lippert and G.M. Meyer, *Einheit auf Befehl?*, Opladen: Westdeutscher Verlag, 1994;

N. Leonhard, 'Armee der Einheit: Zur Integration von NVA-Soldaten in die Bundeswehr', in S.B. Gareis and P. Klein (eds), *Handbuch Militär und Sozialwissenschaft*, Wiesbaden: Verlag für Sozialwissenschaften, 2004, pp. 70–80.
2 R. Geissler, 'Ethnische Minderheiten', *Informationen zur politischen Bildung*, 269, 2000, 32.
3 *Aktuell 2004. Mit aktuellem Länderlexikon*, Dortmund: Harenberg Lexikonverlag, 2003, p. 70.
4 A law issued in 1992 (*Kriegsfolgen-Bereinigungsgesetz*) has introduced the notion 'late resettler' for all the resettlers since 1993.
5 The initially discussed formation of a separate unit only for Muslims, as in Austria, is no longer on the agenda.
6 G. Kümmel and H. Biehl, *Die anderen Soldaten. Spätaussiedler, religiöse Minderheiten und Eingebürgerte in der Bundeswehr*, Strausberg: Sozialwissenschaftliches Institut der Bundeswehr, 2001.
7 Ibid., pp. 10ff.
8 B. Fleckenstein, *Homosexuality and Armed Forces*, Munich: Sozialwissenschaftliches Institut der Bundeswehr (SOWI Working Paper 84), 1993, p. 21.
9 H. Kujat, *Führungshilfe für Vorgesetzte: Umgang mit Sexualität*, Berlin: Bundesminister der Verteidigung FüI4, 2000, p. 4.
10 F.W. Seidler, *Frauen zu den Waffen? Marketenderinnen, Helferinnen, Soldatinnen*, 2nd edn, Bonn: Bernhard & Graefe, 1998.
11 I. Anker, E. Lippert and I. Welcker, *Soldatinnen in der Bundeswehr*, Munich: Sozialwissenschaftliches Institut der Bundeswehr (SOWI Report 59), 1993.
12 Source: German Ministry of Defence, 1999.
13 Anker *et al.*, *Soldatinnen in der Bundeswehr*, p. 112.
14 P. Klein, G. Kümmel and K. Lohmann, *Zwischen Differenz und Gleichheit. Die Öffnung der Bundeswehr für Frauen*, Strausberg: Sozialwissenschaftliches Institut der Bundeswehr (SOWI Report 69), 2000.
15 P.M. Zulehner and R. Volz, *Männer im Aufbruch. Wie Deutschlands Männer sich selbst und wie Frauen sie sehen. Ein Forschungsbericht*, 3rd edn, Ostfildern, Germany: Schwaben-Verlag, 1999; M. Meuser, *Geschlecht und Männlichkeit. Soziologische Theorie und kulturelle Deutungsmuster*, Opladen, Germany: Leske & Budrich, 1998; R.W. Connell, *Der gemachte Mann. Konstruktion und Krise von Männlichkeiten*, Opladen, Germany: Leske & Budrich, 1999.
16 R. Seifert, *Militär, Kultur, Identität. Individualisierung, Geschlechterverhältnisse und die soziale Konstruktion des Soldaten*, Bremen: Edition Temmen, 1996.
17 F. Faludi, *Backlash. Die Männer schlagen zurück*, Reinbek, Germany: Rowohlt, 1995.
18 See also R.M. Kanter, *Men and Women of the Corporation*, New York: Basic Books, 1997.
19 H. Biehl and G. Kümmel, *Warum nicht? Die ambivalente Sicht männlicher Soldaten auf die weitere Öffnung der Bundeswehr für Frauen*, Strausberg, Germany: Sozialwissenschaftliches Institut der Bundeswehr (SOWI Report 71), 2001.
29 G. Kümmel and I.J. Werkner (eds), *Soldat, weiblich, Jahrgang 2001. Sozialwissenschaftliche Begleituntersuchungen zur Integration von Frauen in die Bundeswehr – Erste Befunde*, Strausberg, Germany: Sozialwissenschaftliches Institut der Bundeswehr (SOWI Report 76), 2003.
21 P. Klein, 'Die Streitkräfte – eine nationale Insel', *Information für die Truppe*, 43, 5, 1999, 92.

13 Diversity in the Belgian armed forces

Philippe Manigart

Introduction

The Belgian armed forces, like other complex organizations in the post-modern global world, have become more diverse internally and are also operating in an ever more diverse environment. Their new missions are themselves very diverse (the fight against terrorism, peacekeeping, humanitarian missions, monitoring, etc.), take place all over the world in culturally, ethnically and linguistically very diverse regions, and are conducted most of the time under the guidance of intrinsically ambiguous rules of engagement, in a multinational framework (NATO, Eurocorps, but also all the new so-called task forces, force packages, modular structures, etc.). For example, Belgian troops are presently deployed in Afghanistan as part of ISAF, in Kosovo as part of KFOR, and in Bosnia as part of EUFOR. But in the past few years the same troops, or other units, were part of multinational task forces sent to Somalia, Rwanda, Zaire, Haiti, Cambodia, Turkey, etc. Finally, given the nature of these new missions, the rules of engagement of these forces are intrinsically ambiguous. In other words, as is the case with private companies operating in a global economy, the Belgian armed forces face internal as well as external diversity.

The aim of this chapter is twofold: first, to analyse how diverse the Belgian armed forces are and what attitudes soldiers have towards this increasing diversity; and second, to describe how diversity management is approached in the Belgian armed forces.

How diverse are the Belgian armed forces?

One can distinguish at least five institutionalized dimensions of diversity within the Belgian armed forces. In historical order of appearance, these are language (French- versus Dutch-speaking), gender (men versus women), ethnicity ('Belgians' versus Belgians of non-Belgian descent), sexual orientation (heterosexuals versus homosexuals)[1] and, since 2004, nationality (Belgians versus other EU citizens).

Linguistic diversity

It is important to remember that, like the Canadian and Swiss armed forces (among others), the Belgian armed forces were culturally diverse from the beginning: Flemish- and French-speaking soldiers (and German-speaking ones) have always served in the military. This linguistic diversity, however, has not always been officially recognized. For instance, in October 1834, by regulation, the use of Dutch was prohibited in the armed forces. Although this book is mainly concerned with the 'new' forms of diversity in armed forces, it is not without interest to briefly describe the main steps in the linguistic integration of the Belgian armed forces.[2]

In 1899 the Military Code of Justice was revised to allow the possibility of Dutch-speaking proceedings in the Military Court, which until then had been prohibited, leading to numerous incidents. On 2 July 1913, the first comprehensive law on the use of languages in the military was voted in by Parliament. This law should have been put into practice on 1 January 1917. Among other things, it foresaw that the basic training of draftees had to be given in the two national languages. These measures, however, came too late, because of the start of World War I, and could not prevent the numerous incidents that took place in the trenches in Flanders fields (French-speaking officers giving orders in French to Flemish soldiers who could not understand them).

It was not until 1938 that a law concerning the use of languages in the military passed through Parliament and was put into practice. This law imposed bilingualism for all officers and reinforced the unilingualism of units. In relations with administrative authorities, the language of the region had to be used. So, until the 1930s French was the only official language within the armed forces and officers were all French-speaking (which does not mean that there were no Flemish officers, only that they were not Dutch-speaking). At the Royal Military Academy, French remained the only official language until 1935. In 1939 the proportion of candidates to the Royal Military Academy who were French-speaking was 81 per cent. Not until the 1950s did the proportion of Dutch-speaking candidates begin to reflect their proportion in the population.[3]

These days the use of languages in the military is even more regulated than it is in the civil service. For example, military regulations stipulate that a superior must always speak to a subordinate in the subordinate's language.

After the Second World War a quota system was introduced at all levels of the hierarchy, so that now the Belgian armed forces are roughly representative in terms of language (60 per cent Dutch-speaking versus 40 per cent French-speaking). Until the end of the twentieth century, French- and Dutch-speaking personnel served, for the most part (except officers), in segregated (i.e. unilingual) units. For example, there were (and still are) a French-speaking Army brigade (in Marche-en-Famenne) and F-16 Air Force wing (in Florennes) on the one hand and a Dutch-speaking brigade (in Leopoldsburg) and F-16 wing (in Kleine-Brogel) on the other hand.

With the end of the Cold War, however, there has been a reversal of this trend. The reason has been a drastic reduction in size and restructuring of the Belgian defence forces, which in 2000 led to the so-called *Strategic Modernisation Plan 2000–2015*.[4] A key aspect of this plan was the adoption of a joint structure, for operational units as well as for support and staff activities, which led to a removal of horizontal boundaries and better integration of civilian personnel into the structures of the armed forces.[5] Consequently, military and civilian personnel who previously had worked in two separate structures and in mostly unilingual units are now integrated into one – mostly bilingual – structure and have to work together more closely than before.

Also, with the multiplication of crisis response operations all over the world, the operational forces (brigades, wings, flotillas, etc.) have adopted a modular structure where the various components are 'packages' to be assembled according to the specific mission. In this context, more and more soldiers from both languages find themselves working alongside each other, as part of the same mission, in the theatre of operations. The quota system, however, has more or less remained in place.[6] Table 13.1 shows the distribution of military personnel by rank and language in 2005. As one can see, the proportion 60 per cent Dutch-speaking versus 40 per cent French-speaking is more or less respected. That said, there are more French-speaking candidates than Dutch-speaking ones and the retention rates of the former tend to be higher, particularly at private soldier level. This phenomenon can be explained, first, by the continuing higher unemployment rate among young people in Wallonia and, second, by the traditionally greater attractiveness of a military career and the better image that the armed forces have in southern Belgium.[7]

As far as German-speaking citizens are concerned, during the draft there was a German-speaking infantry company. This company was dismantled in 1995 as a result of the restructuring of the Belgian armed forces, and since then German-speaking soldiers have served in French-speaking units, which, as a former Defence Minister, Jean-Pol Poncelet, once recognized, is a form of discrimination.

Ethnic and national diversity

When people with different habits and worldviews come together in the workplace, misunderstandings and conflicts inevitably occur as a result of dissimilar

Table 13.1 Personnel distribution by language in the Belgian armed forces, 2005 (%)

Category	Dutch	French
General officers	58	42
Senior officers	61	39
Junior officers	60	40
NCOs	57	43
Privates	51	48
Total	55	45

Source: Ministry of Defence.

expectations and norms. In a certain sense, however, in Belgium as in most European countries, but unlike in the United States this cultural/ethnic dimension is rather new, as European societies have long been considered more homogeneous societies. But this is fast changing: what the Canadians call 'visible minorities' are more numerous than before.

In Belgium, successive reforms of the nationality code (laws of 28 June 1984, 13 June 1991, 6 August 1993 and 1 March 2000) significantly facilitated the acquisition of Belgian nationality.[8] Consequently, the number of Belgians of foreign descent (especially non-Europeans) has been increasing during the past few years. The pool of young Belgians is thus becoming more multicoloured, more ethnically diverse. However, unlike in other countries (such as Great Britain, the Netherlands, Canada or the United States), in Belgium no statistics exist on the number of Belgians of foreign descent, either for the general Belgian population or for the armed forces. By law, ethnic origin cannot be mentioned on any official document or statistics (birth registers, the population census, etc.).[9]

To the extent that 'original Belgians' are not particularly attracted by a military career any more, because it is not very prestigious and, with the multiplication in the number of crisis response operations, it has become a more dangerous career choice, the armed forces are bound to become ethnically more diverse and multicoloured. Indeed, as in most Western countries, the Belgian armed forces constitute an opportunity for upward mobility for low-educated, low-income young people, as well as for those belonging to ethnic minorities. It should immediately be pointed out that by becoming more ethnically diverse, the Belgian armed forces can increase the quality of their personnel. Their recruitment base becomes larger and they are thus able to be more selective in recruitment. In this light, the *Strategic Modernization Plan 2000–2015* called for a more proactive recruitment of men and women from ethnic minorities.[10] This policy was reiterated in February 2003 in the updated version of the plan, the *Strategic Plan +*.[11]

On 30 October 2000, in an interview with the French-speaking newspaper *La Dernière Heure*, André Flahaut, the Belgian Defence Minister, went one step further in the widening of the ethnic diversity in the Belgian armed forces by airing the idea of recruiting EU – and even non-EU – citizens living in Belgium. According to Flahaut, the recruitment of EU citizens would prefigure a future European army, while recruiting non-EU citizens living in Belgium would be a means to integrate them into society. As the recruitment of non-EU citizens would have required some constitutional changes,[12] the government decided to restrict the recruitment of non-nationals to EU citizens only. A new law concerning the recruitment of military personnel was adopted on 27 March 2003 by the Belgian Parliament.[13] Article 8 of this law states that to be able to apply for a job in the military, one has to be Belgian or a citizen of a member state of the European Union. Recruitment began in 2004; as of July 2005, a total of 296 candidates had applied for a job and twenty-one had been recruited. Table 13.2 shows the distribution of foreign recruits by rank and nationality. As one can see, the bulk of the recruits are privates (19), with only one (Dutch) NCO and

Table 13.2 Distribution of foreigners by rank and nationality, 2005 (*N*)

	Recruits	Applicants
Rank		
Officers (cadets)	1	51
NCOs	1	52
Privates	19	193
Nationality		
Dutch	10	74
French	6	76
Italian	1	66
German	1	23
Spanish	1	17
Portuguese	1	14
Polish	1	4
Other	0	22
Total	21	296

Source: Ministry of Defence.

one (French) person who has been admitted to the Polytechnics Faculty of the Royal Military Academy as a candidate pilot. The reason so few have been effectively recruited, especially at officer level, is mainly to do with a lack of proficiency in the French and/or Dutch language.[14] As a result, it is not surprising that most of the recruits (ten and six respectively) come from the Netherlands and France.

It should be pointed out that Belgium is not the only EU country to allow the recruitment of non-nationals: Luxembourg and Spain have recently passed legislation to permit the recruitment of EU citizens and of Spanish-speaking citizens from Latin America, respectively. Ireland, Great Britain and France have been recruiting non- nationals for a long time.[15] This 'denationalization' of the military function is a rather logical evolution. On the one hand, it is the result of the globalization process: recruitment in global business firms is no longer based on nationality criteria. On the other hand, from a purely military viewpoint the nationality criterion as a condition for access to the armed forces dates back to the beginning of the nineteenth century and is linked to the appearance of the concept of the nation-state (and thus of nationalism) and citizenship. The era of the nation-state – especially in Europe – is, however, on the decline. With the advent of the single market, free movement of workers is now a reality. In this context the opening of military organizations to EU citizens appears to be a logical next step, especially when recruitment of nationals has become problematic.

Diversity of gender and sexual orientations

One important dimension of diversity in organizations is gender diversity. The workforce of our post-industrial societies is rapidly moving from being male dominated to a position of equality in numbers between men and women. Nevertheless, there remain many barriers for women seeking equal treatment in most organizations. What is the situation in the Belgian armed forces? And what about the latest dimension, i.e. sexual orientation? These questions will be adressed in this subsection.

Historical background

Among NATO countries, Belgium began to recruit women in her armed forces rather late, i.e. in 1975, the International Year of Women, for enlisted personnel and in 1977 for officers.[16] In 1978 the first female cadets were admitted into the Royal Military Academy. The decision to recruit female personnel was directly linked to a decision taken by the government in 1974 to partly professionalize the military. Facing a shortage of candidates, it was decided to open the doors of the Belgian armed forces to a new population segment, i.e. women.

Until 1981, women did not have access to some functions, namely combat functions.[17] In February 1981 the Defence Minister, against the advice of the Joint Staff, decided to open all functions in the armed forces to women. Since then, therefore, all functions have been open to both men and women, without distinction. Following civilian legislation on equal opportunity employment, job descriptions cannot make reference to gender, and any gender discrimination in hiring, working conditions, admission on to training programmes, etc. is prohibited. There is also no distinction made between peacetime and wartime. In 2003 a new law intended to fight all forms of discrimination extended the definition of 'discrimination' to criteria other than gender, namely race, colour of skin, national or ethnic origin, sexual orientation, religious or philosophical convictions, civil status, age, health, handicap, etc.[18] The law also applies to the military.

The new legislation meant, among other things, that physical tests should be the same for both men and women. Before 1981 there existed specific physical tests for men and women. In response to the ministerial decision, the Joint Staff decided to suppress the female tests and to apply the formerly male tests to everybody. Natually, this decision proved discriminatory, to the extent that these tests were not gender neutral, but favoured men.[19] In order to remedy these discriminatory practices, the Minister of Defence ordered the Joint Staff's Personnel Directorate to develop new, gender-neutral tests. These gender-neutral tests were implemented in 1986,[20] with the result that the number of women increased.[21]

The accession of women to all functions sometimes posed practical problems, however, for example in the Navy. In other navies, ships have been refitted in order to allow women on board. In Belgium, however, no such thing occurred. Basically, as long as there were at least two separate sleeping quarters on a ship, it was decided that women could, in principle, serve on board. So, at the present time, women are allowed on board all types of ships in the Belgian Navy.

Female representation

As Table 13.3 shows, although their number has been increasing, women remain largely under-represented in the Belgian armed forces. In 2005, female personnel represented 8.4 per cent of total personnel, up from 1.2 per cent in 1976.

Up to 2002, the higher the rank, the fewer women there were. Indeed, in 2002, if women composed 10.3 per cent of the total among enlisted personnel (privates), the corresponding percentages were only 6.3 per cent among NCOs and 5.1 per cent among officers. As Table 13.3 makes clear, however, the rate of increase among female officers has been higher than among the other categories, so that in 2005 female officers slightly outnumbered, in relative terms, female NCOs (7 per cent versus 6.7 per cent). At the Royal Military Academy the percentage of female cadets is even higher (around 17 per cent in the past few years),[22] reflecting a trend observable in higher education, where female students now outnumber their male colleagues in almost all faculties, except engineering. As of January 2006, the highest-ranking female officer was a general, who was also the commander of the Medical Component.

The most 'female-friendly' service is the Medical Service, with 21.4 per cent women. The Army, the most 'traditional' service, has the smallest proportion of female personnel (6.7 per cent).

As Table 13.4 shows, women are very under-represented in the combat arms. For instance, in 2005 there were only six female personnel in Infantry (0.4 per cent), thirteen in Armour (1.5 per cent) and six pilots (1.5 per cent) in the Air Force. The highest percentages of female personnel were to be found in the services section in the Navy (30.5 per cent), and in the administration and communications and information systems in the Army (13.2 per cent and 11 per cent respectively).

Sexual orientation

For reasons of privacy rights, there are no statistics available on the number of gays, lesbians and transsexuals in the Belgian armed forces, nor have any

Table 13.3 Distribution of women among categories in the Belgian armed forces, 1976–2005

Year	Officers		NCOs	Enlisted	Total
	N	%			

192 P. Manigart

Table 13.4 Occupational distribution of female officers by force, 2005

Category	Army		Air Force	Navy
	N	%		

surveys ever been carried out on this topic within the Ministry of Defence. All one can say is that the official policy under the present Minister of Defence is the one in force in both public service and the private sector – that is, of accepting and respecting people's diverse sexual orientations. In other words, it is clearly a policy beyond the famous 'don't tell, don't ask' attitude of the US (and other) armed forces.

There are examples of gays and lesbians of all ranks who have publicly 'come out' and affirmed their sexual orientations. There is also at least one case of a transsexual NCO who, while on active duty, underwent surgery (reimbursed by social security) in order to become a woman and who remained in her current position. The head of her service informed her colleagues of her operation in order to smooth her reintegration. At present this person is still working in the same service with the same colleagues, without (apparently) too many problems.

Attitudes towards women and ethnic minorities in the Belgian armed forces

Attitudes on ethnic relations

From a 1998 qualitative survey on the mechanisms that can lead to racist attitudes in the Belgian armed forces,[23] it appears that in the 1990s the Belgian military tended to play down gender and ethnic diversity. Although the study concluded that racism was not more prevalent in the Belgian armed forces than in the parent society, it nevertheless brought to light a number of disturbing facts. Among them were the presence in some units of far-right sympathizers (the report recommended that such individuals not be tolerated in the armed forces) and the use by some instructors of racist language during training.

Overtly racist or xenophobic attitudes were more common among enlisted personnel than among officers. Besides pointing out a general educational and social effect (the less educated one is and the lower one's social class, the more one tends to express overtly racist or xenophobic opinions), one particular explanation was given by the researchers: the formation of officers at the Royal Military Academy pays more attention to the acquisition of social and cultural skills than the training schools for enlisted personnel tend to do.

In a 1998 quantitative survey on the internal identity of the Belgian military, two items dealt with the topic of racism and tolerance.[24] Military respondents were asked whether, in their opinion, the employees of the Belgian armed forces were rather racist or rather tolerant. On a scale from 1 ('racist') to 10 ('tolerant'), the mean was 6.55. In other words, a large majority of respondents (73 per cent) thought that Belgian soldiers were rather tolerant (codes 6 to 10 on the scale). The higher the rank, the less the tendency to think that Belgian military employees were racist: the percentages went from 31 per cent among enlisted personnel (privates) to 28 per cent among NCOs and to 19 per cent among officers. Female employees were somewhat less likely than their male colleagues to think that soldiers were rather tolerant, but the trend was very slight (the Pearson correlation coefficient was equal to 0.07).

Attitudes on gender integration

From a 1990 survey among female and male personnel,[25] it appeared that, if the values of male and female personnel did not differ in a fundamental way, there were nevertheless slight differences of perceptions. Women were a little more occupationally oriented than men. For instance, they were more likely to say that they had enlisted to escape unemployment or to be financially independent. Men, on the other hand, attached more importance to items such as 'because it's a job with plenty of action' or 'to serve my country'.[26] Successive yearly surveys of entrant cadets of the Royal Military Academy tend to confirm these findings, though the differences tend to be less important than in 1990.[27]

The 1998 survey on the internal identity of the Belgian military contained a few questions on male and female attitudes towards women in the military. If 39 per cent (against 50 per cent) of the respondents thought that the forces were a man's world and only 19 per cent (against 81 per cent) that women do not belong in the military, 48 per cent (50 per cent of men versus 27 per cent of women) believed that women do not perform as well as men in combat units.[28]

Finally, and more to the point as far as our topic is concerned, the 1990 survey found that sexual harassment was, at the time, a sensitive problem. Indeed, if 90 per cent of female respondents said they felt well integrated in the Belgian military, that their relations with superiors were generally positive and that their relations with male colleagues were very or rather good (78 per cent, versus 62 per cent for men), a significant number, however, complained about several forms of sexual harassment.[29] For example, 65 per cent said that they were sometimes victims of voyeurism; 75 per cent cited obscene remarks and 54 per cent subtle verbal sexual advances. More serious forms of sexual harassment were also cited by a significant minority of female respondents: exposure to pornographic material (35 per cent), overt sexual advances (28 per cent), pawing (39 per cent). Five per cent even claimed to have been victims of sexual blackmail and 3 per cent of sexual violence.

Unfortunately, because there were no comparable data for other organizations, one cannot say whether these percentages differed significantly from those in the civilian sector, but in any case these were quite disturbing percentages in themselves. A sensitive (and unanswerable) question is whether, by 2005, sexual harassment had become less prevalent than fifteen years earlier. Complaints about sexual harassment by male colleagues were also regularly brought forward in the 1998 qualitative survey, as well as continuing negative attitudes on the part of some male soldiers (especially older ones) towards the presence of women in the military.[30]

Intercultural management and training in the Belgian armed forces

Increased emphasis on work teams within armed forces (as in civilian organizations), increased cultural diversity of the workforce and globalization of the theatre of operations are all making diversity management a high priority. As the term 'managing diversity' implies, the organizational interventions that fall within the realm of this label focus on ensuring that the variety of talents and perspectives that already exist within an organization are utilized well. Against this background, I will now describe the policies of the Belgian military in this domain.

Cultural diversity

Before the 1990s, cultural diversity was not an issue in the Belgian military. In fact, it took the events in Somalia in 1993 to force political and military authorities to confront the problem of cultural diversity and to face the racism and xenophobia within the armed forces in their relations with local populations.

In 1997, after three soldiers accused of violence against Somalis were acquitted by the military court and after other acts committed by Belgian soldiers in Somalia became known, Jean-Pol Poncelet, the Defence Minister at that time, ordered the conduct of a study by the Centre for Equal Opportunities and Opposition to Racism. The study was to focus on the mechanisms that could lead to racist attitudes in the Belgian armed forces and recommend the kind of changes needed to improve the situation. The report, known as the Leman Report from the name of the director of the centre, was submitted to the minister in April 1998. The report recommended the recruiting of more women, and people from ethnic minorities, and the organizing of more contacts with ethnic organizations and communities. Most important was the fact that right from the beginning this respect for diversity should be clearly stated and reaffirmed, for instance in a value statement or as part of a mission statement for the Belgian armed forces.[31]

As far as operations abroad were concerned, it was recommended that training prior to missions be adapted away from the purely technical aspects towards more emphasis on communications and intercultural skills, and information on the local culture and norms of behaviour. Also clear guidelines should be issued on the prohibition of racist behaviours or attitudes within the Belgian armed forces. The report also clearly stated that officers should never condone racism or sexism.[32]

A new General Order (J/827) clearly defining the policy concerning the fight against racism and xenophobia was drafted by the Joint Staff and signed by the minister in February 1999, integrating the recommendations of the Leman Report in this domain. André Flahaut, the Defence Minister who succeeded Jean-Pol Poncelet in 1999, intensified the work of his predecessor in this domain by making diversity, equal opportunities for men and women, and the fight against all forms of discrimination strategic priorities of his policy. These priorities were announced in his *Strategic Plan 2000–2015*[33] and further developed in the *Strategic Plan +*.[34]

Concrete actions – legislative, statutory and organizational – followed. So, in 2003 a *Service d'Inspection Générale* (Inspector General), led by a General Officer, was created, and tasked among other things with equal opportunity policy and with reviewing all forms of complaints in matters of sexual and moral harassment within the Department. In 2004 the Directorate General Human Resources promulgated a general policy document on violence, sexual and moral harassment at work (DGHR-APG-OGWCIT-001). The preamble, section 1 of this document, states that military and civilian personnel of the Department of Defence must act in accordance with the values contained in the Universal Declaration of Human Rights, the Convention for the Protection of Human Rights and Fundamental Freedoms, and in the Constitution, and that each person must be treated with respect and on an equal basis. Section 4 further states that violence and sexual and moral harassment at work are neither condoned nor tolerated. The document defines the attributions of the various actors tasked with the protection of the personnel against violence and sexual and moral harassment at work.

Gender integration

On 14 February 1997 the Council for Equal Opportunities between Men and Women issued a series of recommendations concerning a more developed equal employment opportunity (EEO) policy within the Belgian armed forces.[35] Among other things, it recommended the adoption of an affirmative action policy in order to increase the number of women serving in the Belgian armed forces. In order to make the armed forces more attractive, especially – but not exclusively – to women, the minister decided to align the Ministry of Defence with the employment practices of the other federal departments and to introduce within his department more diversified and flexible working conditions, such as part-time work, career interruptions, longer parental leave, paid leave, etc. All these measures were aimed at easing the difficulties that female employees, especially, encounter while trying to combine their work and family life.

His successor, André Flahaut, went ahead with the implementation within the Ministry of Defence of general EEO directives at the workplace mandated by the European Commission and the Belgian government. The *Strategic Plan 2000–2015* and the *Strategic Plan +* both contained a section on EEO and called for the development of an EEO Plan for Defence. The strategic EEO objectives of the Ministry of Defence were included in a policy paper presented by Laurette Onkelinx, Deputy Prime Minister and Minister of Employment and EEO Policy, to the Council of Ministers and approved on 25 January 2001. These strategic objectives were aimed at increasing the number of women in the armed forces in general, and in the higher ranks in particular.

The minister also negotiated with the unions representing the military and civilian personnel of the Ministry of Defence a series of measures aimed at improving the position and quality of life of women within the department.[36] Among these measures were the introduction of new physical tests in order to improve equal opportunities between men and women in the armed forces; improved parental leave and childcare facilities; an increase in the number of day nurseries; and a promise to negotiate the introduction of new benefits and flexible work schedules.

Conclusion

One could say that until the 1930s the Belgian armed forces were composed mainly of white male soldiers (and French-speaking officers); of white male and female soldiers of Belgian descent from 1975 on; and that since 1985 they have also become slightly more multicoloured. Externally, the Belgian armed forces are operating more and more in an ethnically and culturally diverse environment and, as the incidents in Somalia showed, they sometimes have difficulties coping with such diversity. These difficulties were what led to the 1998 Leman Report on the mechanisms that can give rise to racist and sexist attitudes in the Belgian armed forces.

The main conclusion of the report was that the Belgian military had to become more diverse and more tolerant of diversity, and that intercultural training should become part of the standard training of all soldiers (but especially of the cadre). This required, however, and still requires, a fundamental change in the organizational culture of the Belgian armed forces. In other words, the Belgian armed forces have to become a truly multicultural organization – in the words of Milkovitch and Boudreau,[37] an organization 'where employees with non-traditional backgrounds can contribute and achieve to their fullest potential. Rather than employees assimilating to the dominant culture, these organizations seek to accept and capitalize on employee differences'.

From such a perspective, the presence of non-Belgians in the Belgian military can be viewed as a competitive advantage for what has become the core business of the post-modern Belgian armed forces, i.e. peacekeeping and humanitarian operations. Another competitive advantage, flowing this time from the linguistic diversity, is the better average language proficiency of Belgian – and more particularly Dutch-speaking[38] – personnel, especially officers, compared to most of their compatriots.[39]

Notes

1 For more details on this dimension in the Belgian armed forces, see P. Manigart, 'Homosexuals and the Belgian Military', in G. Harries-Jenkins (ed.), *Homosexuals in European Armed Forces: Policies, Practices and Problems*, London: United States Army, European Research Office, 1996.
2 Source: 'L'Admission progressive de la langue néerlandaise dans les textes officiels en Belgique', Parl. Doc. Senate, session 1966–67, no. 114, pp. 4–14.
3 V. Werner, 'La Position de l'Armée belge dans le cadre de la société belge', Brussels: Institute of Sociology of the Free University of Brussels, unpublished paper, 1974, p. 19.
4 A. Flahaut, *Le Plan stratégique pour la modernisation de l'Armée belge 2000–2015. Des propositions concrètes pour entrer dans le XXIème siècle*, Brussels: Ministry of Defence, 2000.
5 A. Flahaut, *Plan stratégique +. Evaluation et perspectives*, Brussels: Ministry of Defence, 2003, p. 59.
6 Faced with a chronic shortage of Dutch-speaking candidates and therefore with a number of unfilled slots overall, the Minister of Defence, in 2000 in an interview with a Dutch-speaking newspaper, alluded to the possibility of ending the linguistic quotas and of substituting French-speaking recruits for Dutch-speaking ones, thus in effect making the Belgian armed forces more French-speaking. The idea, however, was never officially put into practice, and at present it is difficult to imagine that a dominantly French-speaking military would be viable on a long-term basis.
7 See P. Manigart, *Les Forces armées belges en transition: une analyse sociologique*, Brussels: Institute of Sociology Press, Free University of Brussels, and P. Manigart, N. Wauters and P.A. Charrault, 'Étude de marché sur le recrutement de personnel volontaire et de carrière dans les Forces Armées belges', technical report, Brussels: Royal Military Academy, Department of Sociology, 1994.
8 In January 2005, Belgium counted 860,287 foreigners out of a total population of 10,396,421 residents, which means that 8.2 per cent of the total population did not have Belgian citizenship. Source: National Institute for Statistics (statbel.fgov.be).
9 The same goes for France, where the Constitution prohibits the mention of the race of a person on his or her identity card or in the census.
10 Flahaut, *Le Plan stratégique*, p. 68.
11 Flahaut, *Plan stratégique +*, p. 61.
12 The Belgian Constitution reserves military jobs to Belgian citizens, with exceptions provided for by law. Recruitment of EU citizens is such an exception, based on the principle of free movement of workers within the European Union.
13 Loi du 27 mars 2003 relative au recrutement des militaires et au statut des musiciens militaires et modifiant diverses lois applicables au personnel de la Défense, Brussels: *Moniteur belge* of 30 April 2003.
14 For officer positions, candidates must pass an exam in both languages.
15 For more details on the conditions, see A. Dumoulin, P. Manigart and W. Struys, *La Belgique et la politique européenne de sécurité et de défense*, Brussels: Bruyland, 2003, pp. 382–384.

16 For more details on the process of integration of women in the Belgian armed forces, see P. Manigart, 'Le Personnel féminin dans les Forces armées belges', *Courrier hebdomadaire du CRISP* 1060, 1984.
17 It should, however, be pointed out that in 1976, at the initiative of Parliament, the explicit reference to *combat* function was suppressed and replaced by 'dangerous and hazardous functions', a vaguer notion.
18 Art. 2 § ^{er} de la Loi du 25 février 2003 tendant àlutter contre la discrimination et modifiant la loi du 15 février 1993 créant un centre pour l'égalité des chances et la lutte contre le racisme, *Moniteur Belge*, 17 March 2003.
19 For example, the number of female cadets at the Royal Military Academy declined drastically after 1981. The best-known case – related by the press at the time – was that of Carine Verbauwen, a Belgian swimming champion, who was not admitted because she had failed her physical tests. Before 1981 there existed a quota system reserving a certain number of places for female candidates.
20 Law of 19 December 1986 modifying the criteria for physical tests.
21 For example, during the past few years female candidates have accounted for almost 20 per cent of new entrants at the Royal Military Academy.
22 Source: Royal Military Academy, Department of Behavioural Sciences.
23 This survey was conducted at the request of the Belgian Defence Minister by the Centre for Equal Opportunities and Opposition to Racism. Its aim was to study the mechanisms that can lead to racist attitudes in the Belgian armed forces. It consisted of eighteen focus groups ($N=180$) and in-depth interviews of 520 individuals. The focus groups were conducted by Research International and the depth interviews by a team from the Rijksuniversiteit van Gent and the Université Libre de Bruxelles. For more details, see J. Leman, *Étude des mécanismes pouvant mener à des attitudes de racisme au sein de l'armée belge*, Report to the Minister of Defence, Brussels: Ministry of Defence, 1998.
24 J. van den Bulck and S. Eggermont, *De identiteit van de Belgische Krijgsmacht Anno '98*, Leuven: Centrum voor Publieksonderzoek, Catholic University of Leuven, technical report, 1998.
25 This survey was also conducted at the request of the Belgian Defence Minister by the Behavioural Sciences Department of the Royal Military Academy among a representative sample of 765 female and 658 male personnel of the Belgian armed forces. For more details, see E. Bauwens, 'Vrouw in de Belgische Krijgsmacht', *Maatschappij en Krijgsmacht* 4, 1990, 22–28; E. Bauwens and J. Van de Vijver (1990), 'Vrouwen en mannen, samen in de krijgsmacht: un défi pour la gestion des ressources humaines', paper presented at the conference on *Des femmes en uniforme: armée, gendarmerie et police*, Brussels, 6 May 1990; K. De Decker and G. Nijs, 'Vrouwen in de Belgische strijdkrachten. Verslag van de opiniepeiling: motivaties, sociale afkomst ... van vrouwelijke militairen', paper presented at the conference *Des femmes en uniforme: armée, gendarmerie et police*, Brussels, 6 May 1990.
26 E. Bauwens, 'Integration of Female Personnel in the Belgian Armed Forces', paper presented at the 1991 Biennial Conference of the Inter-University Seminar on Armed Forces and Society, Baltimore, 11–13 October, p. 4.
27 Source: Department of Behavioural Sciences, Royal Military Academy.
28 Van den Bulck and Eggermont, *De identiteit van de Belgische Krijgsmacht*, p. 46.
29 De Decker and Nys, 'Vrouwen in de Belgische strijdkrachten', p. 19.
30 Leman, *Étude des mécanismes*, pp. 106–112.
31 Ibid., p. 73.
32 Ibid., p. 174.
33 Flahaut, *Le Plan stratégique*, p. 68.
34 *Plan stratégique +*, pp. 60–61.
35 *Advies nr 13 van Februari 1997 van het Bureau van de Raad van de Gelijke Kansen voor Mannen en Vrouwen Betreffende de Vrouwen in de Krijgsmacht*.

36 On this topic, see also P. Manigart, D. Resteigne and R. Sabbe, 'Military Unionism in Belgium', in R. Bartle and L. Heinecken (eds), *Military Unionism in the Post-Cold War Era*, London: Routledge, 2006.
37 G.T. Milkovich and J.W. Boudreau, *Human Resource Management*, Boston: Irwin, 1994, p. 70.
38 For largely historical reasons (French was once the lingua franca of the world and is still an official language in most international institutions), French-speaking Belgians – like French people – tend be less language proficient than Dutch-speaking people.
39 In a multinational survey carried out by the Department of Behavioural Sciences of the Royal Military Academy among the personnel of the Eurocorps headquarters in Strasbourg in 1997, it appeared that Belgians less often than the others said that the use of several languages (at the time French and German were the two working languages within the Eurocorps) was a problem for them. See P. Manigart and J. Van Bladel, *La Satisfaction des personnels au sein du Corps européen: l'enquête multinationale*, Brussels: École Royale Militaire, Chaire des Sciences Sociales, rapport technique SS26, November 1997, p. 114.

14 Diversity in the Dutch armed forces

Rudy Richardson, Jolanda Bosch and René Moelker

Introduction

In the concept policy letter 2007 the State Secretary for Defence demonstrates the policy importance of the integration of ethnic minorities in the Netherlands armed forces. Three years earlier he had stressed the importance of integrating women in the armed services. These efforts illustrate that the highest priority has been given to the recruitment and integration of minority groups in the Dutch armed services over the past three years. However, the Dutch armed services' attention to the integration of minority groups is rooted in the 1980s, especially regarding the integration of women.[1]

As well as women, homosexuals received much attention from the Dutch armed services in the last two decades of the twentieth century. In 1992, for the first time a large survey was conducted on the integration of homosexual servicemen, showing that some two-thirds of personnel kept their distance from gay colleagues. The overall opinion was that gays met with considerable difficulties, hindering their functioning in the Dutch armed services.[2] According to another large survey in 1999, this 'gay-unfriendly' climate had changed into a more 'gay-friendly' climate.[3] This study showed that the distance between gays and heterosexuals had significantly decreased: 91 per cent of the male servicemen showed a low degree of homophobia and 61 per cent of homosexuals felt accepted within the military. In 2006 a third large survey on the integration of homosexuals in the Dutch armed services will be conducted and, as with the other surveys, the Dutch armed services' policy-makers will try to translate the results into concrete measures to accelerate the integration of homosexuals in the services.

The attention paid to minority groups in the Dutch armed services has become increasingly important as a result of changes in the mid-1990s, after the East–West détente. One major change was that conscription ended. The defence organization became an all-volunteer force. In this context the Dutch armed services' policy-makers became interested in the working of the labour market. It instantly became clear that the volume of the workforce with ethnic backgrounds, especially Dutch Turks and Dutch Moroccans, and of women, was growing rapidly.[4] However, these groups were under-represented in the armed services.

The expectation that the new armed services should reflect society in terms of ethnic and female representation became a major policy topic at that time.[5] It became clear that the future workforce of the services could not reach its desired strength without recruiting more and more women as well as more candidates from ethnic minority groups.[6] As a result, new efforts and policy measures to recruit and integrate women and members of ethnic minorities into the Dutch armed services have become important issues, partly because studies show that the need to fully incorporate these groups into the services' workforce has become even more important in the twenty-first century.[7] Therefore, we will describe new developments in the ethnic and gender dimensions of diversity in the Dutch armed services since 2000. First, we shall take a closer look at the developments in ethnic cultural diversity at the policy level and at the level of the armed services organizational culture. Thereafter, the same levels will be described for the gender dimension. These developments will be reflected upon in terms of the consequences for diversity and diversity policy in the Dutch armed services in the future.

Ethnic minorities in the Dutch armed forces

The context

According to recent figures, 1,558,000 people from ethnic minorities live in the Netherlands, which is approximately 10 per cent of the total Dutch population.[8] Dutch policy on ethnic minorities was first formulated in the early 1980s, targeting ethnic groups stemming from Surinam and the Netherlands Antilles and Aruba on the one hand, and those coming from Morocco and Turkey on the other hand: respectively, decolonization compatriots and second-generation immigrants. At the same time, demographic prognoses were already making it clear that these ethnic groups were going to constitute a growing part of the Dutch population, as an effect of family reunion and birth rate. The latter was expected to be higher than for the indigenous Dutch population. Quite tellingly, 50 per cent of those aged under 16 in the most urbanized part of the Netherlands, the Amsterdam–Rotterdam–The Hague region, will stem from ethnic minority groups in the near future. The numbers of people belonging to all ethnic minorities in the Netherlands are shown in Table 14.1.[9]

Dealing with this expanding multicultural character in 1999, Richardson and Bosch[10] wrote that Dutch society has shown relatively few extremist tendencies or outbreaks. Yet even then an awareness had grown that some multicultural ideals were more and more out of touch with real developments. The integration of ethnic minorities was hampered by all kinds of cultural, economic and educational bottlenecks. This applied not only to Dutch Turks and Dutch Moroccans but, among others, also to people from the Dutch Antilles. In a number of cities, members of this latter group became the centre of street-level tensions that have necessitated severe policy measures against their growing criminality, as well as policies aimed at improving their weak social position.

Table 14.1 Division of ethnic minorities in the Netherlands

Ethnic background	Total	%
Turkish-Dutch	331,000	21.2
Surinam-Dutch	315,000	20.2
Moroccan-Dutch	284,000	18.4
Antillean-Dutch	125,000	8.0
African-Dutch	95,000	6.1
Asian-Dutch	81,000	5.2
Iraqi-Dutch	41,000	2.6
Afghani-Dutch	31,000	2.0
Iranian-Dutch	27,000	1.7
Pakistani-Dutch	17,000	1.1
Vietnamese-Dutch	16,000	1.0
Indian-Dutch	13,000	0.8
Brazilian-Dutch	10,000	0.6
Rest	172,000	11.0
Total	1,558,000	100

Since 11 September 2001, not only have inter-ethnic tensions become more pervasive, but manifestations of extremism have entered Dutch society as well. The murder of filmmaker and columnist Theo van Gogh on 2 November 2004 by a Dutch-Moroccan Islamic fundamentalist was the dramatic culmination of this extremism. The murder was followed by a month of severe tensions in Dutch society between Islamic groups and native Dutch – notwithstanding the fact that majorities on both sides behaved and talked moderately. Security measures had to be taken to protect Dutch politicians who received death threats. In particular, the popular and very outspoken parliamentarian Ayaan Hirshi Ali, a strong critic of Islam, like Theo van Gogh, became the target of hate and threats. The prosecution and conviction of members of the so-called Hofstad group has emphasized the (potentially) violent climate.[11]

All this has further changed political perceptions of the multicultural society and has led to a polarization between the rhetorics of hardliners and moderates on topics such as integration, assimilation and citizenship. At the same time, youngsters in particular among ethnic minority groups take strong exception to stereotyping about Islam as a 'backward religion', justifying extremism and crime. In such a climate, local street-level tensions can escalate to national debates. And let us not forget the social-economic side: research has made it clear that policies with regard to the 'ethnic minority labour market' have failed during the past five years, not least because of the unwillingness of native Dutch employers to recruit personnel from ethnic minorities. Not only the labour market but also the domains of education, marriage and income show increased segregation between ethnic minority groups and the majority of native Dutch.[12]

The cultural and economic context of inter-ethnic relations is important for the armed services. The armed forces are an important stakeholder because, as we argued earlier, their future personnel will be recruited partly from these ethnic minority groups, especially from the youngest generation. The next section will explore the topic of recruitment in greater detail.

Recruitment and registration

Recruitment and registration featured in the Dutch armed services' policy on ethnic minorities from 1998 to 2005. The registration was introduced in 1998 by the '*Wet Samen*' ['Together Law'], a Dutch government law for all government departments including the Ministry of Defence, to commit to the registration of the number of employees from ethnic minorities within their labour force and to stimulate integration.[13] Moreover, recruitment policy was intensified as a result of an important piece of labour market research initiated by the department in 2000. The outcomes of this research showed that the Dutch-Turkish group in the sample were most interested in participating in the services, especially the Army (22 per cent), that Dutch Antilleans within the sample show interest in military police work (11 per cent) and that rather few Dutch Moroccans within the sample show any interest in the Dutch forces in general (2 per cent).[14] As a result of these outcomes, the Ministry of Defence concentrated on the recruitment of the Dutch Turkish. A special policy measure, the Commutation Settlement of Turkish Conscription, was introduced to facilitate the recruitment of this group.[15] As yet, it is not clear whether or not this measure has significantly affected the recruitment of Dutch-Turkish youngsters in the armed services.

Furthermore, the ministry became aware that a policy of 'sit back and wait' would not be very successful in recruiting new personnel. Active recruitment via job market meetings, pop concerts and television advertisement clips and also the introduction of a real 'armed forces soap' on television were intended to capture the attention of ethnic minority groups. Since 2000 a large group of 'recruitment scouts' have been part of the services. However, the results of these exertions has not resulted in a higher representation of ethnic minorities within the services, as shown in Table 14.2.

The figures illustrate that the minimal growth of ethnic minorities employed by the forces has come to a stop. This is even more painful because these figures are also 'polluted' with an unknown number of employees with an Indo-European background – that is, first- and second-generation employees from the

Table 14.2 Ethnic minority groups within the Dutch armed forces (%)

	1999	2000	2001	2002
Navy	6.4	6.0	5.4	5.6
Army	8.9	9.4	9.3	8.9
Air Force	6.0	6.1	6.5	6.1
Marechaussee	6.5	7.0	7.3	8.0
Average	7.0	7.1	7.1	7.2

Source: Ministry of Defence.

former Dutch East Indies who have migrated to the Netherlands since 1949, and can be considered 'integrated' as a result of participation in the forces over the past four decades.

The decision to reduce active recruitment efforts within the Ministry of Defence in 2003 is a remarkable one if we consider these figures. Serious reductions in the budget of the Ministry of Defence, the decision of the Dutch government to stop the '*Wet Samen*' in 2003 and some major organizational changes at the level of the ministry itself have resulted in less policy concern with active recruitment and integration of ethnic minorities. The Army and the Air Force decided to execute their own policy, mainly because of the threat of a shortage of personnel in general. Army policy-makers, for example, participate in city council panels to discuss with representatives of ethnic minority groups, local politicians, labour market experts and representatives from civil society organizations the possibilities of increasing the participation of ethnic minority groups in the labour market. The Air Force formulated a policy document with guidelines to facilitate the recruitment of people from ethnic minority groups.[16]

As a result of questions in the different services on how to execute ethnic minority policy in the future and as a result of negative outcomes of evaluation research of the developments of ethnic minority policy from 2000 to 2005, the central staff of the Ministry of Defence has decided to strengthen its influence on ethnic minority policy and to reformulate central guidelines to facilitate recruitment and integration.[17]

Organizational culture

Even though ethnic minority policy in the years 1998–2005 was mainly directed at recruitment and registration, some efforts to change the organizational culture in this context were apparent as well. The Ministry of Defence became aware that tensions in society might be reflected in the defence organization and that preventive measures had to be taken. The first signs of tensions had already been visible in early research, showing that 40 per cent of those with a Surinamese ethnic background felt non-integrated and pointing at the 'integration climate' as the most important factor in failed integration.[18]

In more recent years, anecdotal evidence has seemed to confirm these findings. For instance, a physician in the Army, an ethnic minority lieutenant-colonel, was constantly confronted with stories of undesired behaviour and teasing against ethnic minorities within the army. Also, in 2000 a major of the Air Mobile Brigade resigned because he could not cope with efforts to cover up severe tensions between ethnic minority groups and native Dutch within the brigade and the lack of measures against those tensions. In 2003 a Turkish private was severely threatened and harassed by his colleagues in a school battalion, and in 2004 and 2005 tensions between Islamic soldiers and Dutch natives within the Dutch military police detachment during the SFIR missions in Iraq led to the establishment of a special commission of inquiry.

In this context the Ministry of Defence decided to focus on training and education, thus creating the possibility of discussing differences between the perceptions and norms and values of ethnic minority groups and native Dutch respectively. At private soldier, NCO and junior officer level, military servicemen are confronted with training programmes such as 'becoming familiar with other cultures' and 'respectful interaction with ethnic minorities'. Also, important progress was made in training programmes with subjects such as 'social integrity' and 'ethical dilemmas'. It is the ministry's policy to expand these training programmes in time and for all ranks in the near future.

Moreover, the department officially acknowledged the foundation of the Multicultural Defence Network in 2003. This network, supported by different ethnic cultural groups and by native Dutch, is a formal interlocutor for the Ministry of Defence in the discussion on ethnic minority policy in the present and future. The goal of this network is to facilitate interactions between ethnic minority groups and native Dutch within the armed services and to establish a climate wherein everybody, *irrespective* of their origin, can serve optimally. This means that the focus is aimed at creating an organizational culture of non-discrimination and mutual respect in everyday face-to-face interaction by means of education and training for all ranks, in understanding different cultures and religious practices. Also, the importance is stressed of having Army imams who are qualified to counsel the Muslim minority in the military on religious and ethical issues. As might be expected, a major theme at the moment is reflection and debate on the consequences for the armed services of societal tensions between Islamic and native Dutch.

Notwithstanding all these efforts, the question remains whether the organizational culture really has changed over the past five years. A recent study shows that there has been no significant positive change in the attitudes towards ethnic minorities and their integration within the services.[19] While in 1999 armed forces employees showed positive multicultural attitudes (i.e. positive perceptions on ethnic minorities) and positive attitudes towards diversity policies and their effectiveness, in 2005 overall they showed a 'negative' to 'neutral' attitude towards multiculturalism. However, there are differences in outlook between different categories of people. Better-educated employees, majors through to colonels, female employees and, in particular, members of ethnic minorities lean more towards a positive attitude concerning multiculturalism, while generals, privates and less well educated employees especially show negative attitudes.

Attitudes towards diversity policy measurements and its effects remain positive. With respect to specific acculturation attitudes,[20] the outcomes indicated that servicemen and servicewomen are predominantly inclined towards *assimilation* of ethnic minorities in the public domain, while they are somewhat more reserved in the private domain, in which the majority prefer a *separation* strategy.[21] Regarding attitudes towards multiculturalism, it seems that these attitudes became more negative over the past five years; however, attitudes towards policy measurements and its effects remain positive. The overall conclusion is that the greatest challenge of the Ministry of Defence in the case of ethnic minorities lies at the level of changing attitudes towards ethic minorities in the context of a society in which multicultural tensions have become part of everyday life.

Women in the Dutch armed services

Context

In wider Dutch society, in 2005 roughly 59 per cent of women worked for at least twelve hours a week, against 46 per cent in 1996. So, there is an upward trend, and compared to other European countries the most recent percentage is reasonably high. However, these seemingly positive developments regarding female labour participation do not tell the whole story. This is because the Netherlands ranks number one in part-time work: 70 per cent of the female workforce are employed in part-time jobs instead of choosing a full-time career.[22]

Apparently, in one of the most liberal and emancipated of European countries, mechanisms are at work that greatly influence the position of women in society and foster the continuation of a traditional cultural pattern. Quite a few women still conform to ascribed gender roles by choosing to remain at home. Moreover, the notorious glass ceiling is not easily broken when many women are involved in part-time work. Making a career is difficult when working on a part-time basis. In 2003, only 5 per cent of the boards of directors of the twenty-five largest enterprises were female.[23] This traditional pattern in broader society is mirrored in the armed services. Before going into detail about present-day developments with regard to women in the military, we will give a very brief historical sketch.

On 25 April 1944 a Military Women's Corps was founded.[24] Until the mid-1980s, female soldiers worked in their own 'safe area' within the Dutch armed forces, first in the Women's Assistance Corps and later in organizations known as MARVA (Navy), MILVA (Army) and LUVA (Air Force). These women were never in combat units. In those days no women were allowed to enter military academies.

Legislation and regulations forced the armed forces to open up. In 1952 the Netherlands signed the treaty on the rights of women in New York, which was ratified in 1971. In 1978, women were given access to almost all military educational institutes and training centres, including the Royal Military Academy. The separate Women's Corps was disbanded in 1982, and in 1983 the first women entered the Naval Academy, completing the opening up of the military. At present the status of women in the armed forces is formally equal to that of men. With the exception of the Marine Corps and the submarine fleet, from which women are excluded, they have the same job opportunities and fall under the same regulations.

Even though formally women enjoy equal status, it is very difficult for them to live up to the requirements that are in force in the infantry and in the other combat arms. Nevertheless, gradually the first cohort of women who entered the

Royal Military Academy twenty-seven years ago are reaching the top. In the autumn of 2005, Leanne van den Hoek was appointed as commander of the personnel branch of the Army, thus becoming the first female general in the Dutch military. Furthermore, research into the possibility of integrating women on board submarines was started, also in autumn 2005. In the spring of 2006, after the results of this research had been considered, it was decided not to change current policies – that is, not to allow women into the submarine fleet.

Recruitment and retention

The first difficulty in integrating women in the armed services is getting them in. Therefore, we will address the issue of recruitment first, then we will deal with the question of how to keep women in – that is, the issue of retention. With regard to the latter point, we will discuss the kinds of policies meant to ameliorate the position of women in the armed forces.

Iskra et al.[25] hypothesize that armed forces that have switched to all-volunteer forces will open up to female service personnel. A study by Carreiras[26] demonstrates the importance of conscription rates for the percentage of women in armed forces in NATO countries. The two phenomena are correlated ($R = -0.72$; sig. 0.01). In the Netherlands the issue of recruiting women became a relevant policy topic only after the all-volunteer force had been established, in 1996. As a consequence, the armed forces had to comply with the laws of the labour market, had to change their recruitment system, and had to try to persuade more women to join up.

As Table 14.3 shows, over the past five years the percentage of women in the

Table 14.3 Organizational presence of female military personnel (%)

	2000	2001	2002	2003	2004	2005
General Staff	5.1	6.7	7.6	7.1	7.5	6.9
Joint Support Services	11.6	12.7	12.5	13.9	16.4	15.9
Army	7.7	8.5	8.5	8.5	8.2	8.0
Air Force	8.6	9.4	9.7	9.7	9.6	9.5
Navy	10.3	10.0	10.0	10.7	10.5	10.9
Marechaussee	9.2	9.3	10.2	10.5	10.7	11.1
Total: soldiers	14.9	14.8	13.9	13.7	13.4	12.9
NCOs	4.2	4.9	5.6	6.5	6.6	6.8
Subalt. COs	9.0	9.7	10.1	10.7	11.5	11.6
Higher COs	2.8	3.3	3.8	4.1	4.4	5.1
Generals	0.0	0.0	0.0	0.0	0.0	1.1
Grand Total	8.7	9.2	9.3	9.6	9.6	9.5

Source: Ministry of Defence.

Note
The percentages slightly differ from earlier publications by the MOD. The most probable reason is the new structuring of the Armed Forces. (The total number of military servicemen and women serving in the AF is c.50,000)

armed forces has been rising only slowly to 9.5 per cent. (Back in 1992 it was just 5.2 per cent.) However, during the past three years the influx of women has stagnated. The number of females in operational branches has dropped significantly. Most women are employed in the joint support services. The policy objective is to reach the 12 per cent level by 2010 and, moreover, to have 6 per cent of females at the rank of major or higher, and 3 per cent at the rank of colonel and higher.

Until recently, as in other countries, women were distributed very unevenly across the ranks. In 2003 the situation appeared to have improved slightly, the Netherlands armed forces counting six female colonels. Between 2000 and 2005 the number of senior female commanding officers (with the rank of major or higher) doubled. At present, forty-one serve in the Army, forty-nine in the Air Force and forty-two in the Navy. Only five serve in the Marechaussee, or military police. Even when the size of the the Marechaussee is taken into account – it is the smallest of the four services – senior female commanding officers are under-represented in it. Comparing its organizational size to that of the Navy, for instance, the Marechaussee should have about ten female commanding officers. Overall, if women were proportionally represented at the rank of general, there should be ten female generals in the military as a whole.

While women are under-represented at the rank of major and higher, they are over-represented in the short-term contracts and in the lower ranks. This phenomenon is related not only to the recruitment issue but also to the question of retention. It appears that female service personnel quit the organization in relatively large numbers at the age of 30–35, wanting to establish a family without the burden of frequent deployments.

For over ten years, deployments have been core business of the Netherlands armed forces. Seven per cent of personnel deployed in 2003 were female. In absolute numbers, 574 female and 8,093 male soldiers were deployed. The UN resolution 1325 is intended to reinforce the role of women in conflict resolution. The Dutch armed forces comply with this resolution and integrate gender perspectives into peacekeeping operations. The objective is to deploy female soldiers in peace support operation because of the constructive roles they can play.[27] Conflict resolution and the reconstruction of societies will benefit from these policies, and that is yet another reason why the recruitment and retention of female service personnel is an important policy issue.

As Table 14.4 makes clear, in 2004 only 13 per cent of all female soldiers in the Army were in combat units, whereas the percentages of females in combat functions in the Navy and the Air Force are much higher (33 per cent and 21 per cent respectively). Moreover, women are poorly represented in technical units. The reverse is true for support units, in which women are very well represented (c.70 per cent).

All these trends – uneven distribution in the ranks as well as in the services and the pattern of mid-career quitting because of family reasons – have to be taken into account when policies to enhance recruiting and retention are being developed. State Secretary Van der Knaap introduced the policy objective of

Table 14.4 Percentage of women by occupation in the Dutch armed forces in 2004 (only female service personnel, civilians are excluded)

Function/branch	Army		Navy	Air Force	Total
	N	%			

recruiting 30 per cent female personnel by 2010. In 2003 the armed forces succeeded in recruiting 11 per cent female personnel. Time will tell whether or not the objective of increasing recruitment to 30 per cent is feasible.

There is certainly no lack of policy measures especially intended to retain women. The General Equal Treatment Act 1994 and the Working Conditions Act are strong devices to improve equality and working conditions; conduct unbecoming and sexual intimidation are forbidden and punishable. Also, gender ambassadors, confidential councillors and equal treatment committees help in dealing with complaints. Unfortunately, these measures have not prevented the occurrence of sexual harassment, sometimes of a very grave nature.[28] These kinds of incidents illustrate the importance of sound legislation and regulation, leadership that does not accept violations of personal integrity, and culture change.

Measures such as childcare facilities and maternity and paternity leave are also meant to improve the labour participation of women. As a special arrangement for women, the right of re-entry for up to six years after leaving the service has been established. Women also have a right to be exempted from deployment in peace support operations and naval exercises if they have children up to the age of 4. The armed services are also looking into the possibilities for in-house childcare, i.e. at or near the workplace; a number of pilot projects have been started. The government contributes to the cost of childcare. New legislation (in effect as of January 2005) expects a contribution of one-third from employers, even though this contribution is not obligatory.

All in all, policies, policy objectives and regulations seem well in order in the Netherlands. However, in November 2005 the Social Democratic Party, in collaboration with one of the military unions, sounded the alarm. Despite all good intentions and good plans, it was argued, results have been disappointing. As we have already discussed (see Table 14.3), the percentage of women is stagnating or even falling. Also, many women drop out of basic military training prematurely. All of this makes the realization of policy objectives with regard to recruitment and retention seem unlikely.

Therefore, a Social Democratic member of Parliament, Angelien Eijsink, has put forward a number of recommendations to amend ongoing policies.[29] One of the most important recommendations is to separate the young and vulnerable recruits during basic military training and educate them in single-sex platoons. In this way the female soldiers are relieved from being tokens in the first few months of their career, which should improve attrition rates. Eijsink's ideas also aim at appointing more female generals, including in services other than the Army, in order to create positive role models. Whether these recommendations will be included in the official personnel policies is not clear. And whether they would have the intended effect is far from certain. One way or another, as we will argue in the next section, these kinds of policy measures are not enough to change the culture of the armed forces.

Organizational culture

Regulations, legislation, informal networking and policy objectives all help to improve the position of women in the armed forces. Yet they are only preconditions, for it is difficult to change norms and values. Much resistance to integration is – almost – invisible, embedded in masculine culture that tends to exclude women.[30] The strongest resistance is found in traditionally masculine branches such as the infantry.[31] Culture change may be promoted by 'diversity training', but socialization into the masculine culture is much stronger. More effective and efficient for changing the organizational culture is demographic management. Women will rise to equal status only when they are represented in larger numbers and in higher ranks.

The cultural barrier can also be overcome by effective networking. For thirteen years the Defence Women's Network (Defensie Vrouwen Network, DVN) has been a change agent that tries to influence policy-makers and to promote the interests of women in the armed forces. The Defence Women's Network consists of active female service personnel.[32] The network operates as an equivalent of the 'old boy network'. It is difficult for a female NCO to ask for direct help or advice from other females who may or may not be higher in rank. Thanks to meetings of the network, the discussions and the newsletter, it is easier for women in the military to establish informal contacts with other female network members. They can help each other because they form an informal 'old girl' network. Official objectives of the network are to inspire, stimulate, inform and motivate women who are employed in the Dutch armed forces. The network wants to strengthen the position of women and stimulate their advancement to higher positions within the military and the Ministry of Defence.

Regarding gender policies in general, Carreiras[33] concludes that

> if military men feel overly pressured by institutional policies or these are interpreted by both men and women as sources of inequity, blatant resistance to women's integration 'may fade only to be replaced by more subtle, covert forms of discrimination and hostility'.[34] ... policies may be a necessary but not sufficient condition to ensure the sustainability of the process of gender integration in the military; greater inclusiveness in women's military

participation will probably depend on change regarding women's 'controlling' presence in society, its impact on cultural conceptions of gender relations and on a more balanced distribution of domestic and paid work between the sexes.

Conclusion

We have presented the developments with regard to diversity issues in the Dutch armed forces along the dimensions of ethnicity and gender. The first trend is that, over the years, the Dutch armed forces' policy has stressed deficiency, and emancipation of both women and ethnic minorities.[35] This resulted in the Emancipation Letter of 1997 and the policy measurements on gender in 2000 and later. Policies regarding ethnic minorities were introduced after 1998, encompassing initiatives and measures to recruit ethnic minorities and aiming to enhance the position of ethnic minorities in the forces. Both the Defence Women's Network and the Multicultural Defence Network have discussed the existence of 'old boy networks', glass ceiling problems in management development, and recruitment systems.

However, more than emancipation alone is needed. Granted, emancipation has to do with overcoming inequalities and improving the backward position of minorities, and it is important because it may lead to equal treatment and equal opportunities. But does emancipation really change the diversity climate? Unfortunately, cultural barriers largely remain the same, in spite of improvements in training and education in mutual respect and non-discrimination. On the surface it seems that the emancipation of women in society has become more and more accepted, but diversity climate research shows that the majority of white male employees still demonstrate negative attitudes towards minorities, especially when minority policies lead to special treatment in recruitment and facilities ('positive discrimination').[36] Therefore, emancipation will not be efficient when it lacks 'cultural empathy' with others who have different values, norms and perceptions.

Nine-eleven and its aftermath, as well as the murder of Theo van Gogh, have led to serious criticism of the traditional Dutch attitude of 'cultural empathy' and 'cultural relativism' towards ethnic minorities. In this context, Welsch refers to 'real diversity' and introduces the concept of *transculturality* as a reaction on the traditional concepts of culture and multiculture.[37] Transculturality refers to the mixes and permeations in present-day cultures on both the macro (society) and the micro (individual) level, society and the human as 'cultural hybrids'. Future discussions on the integration of minorities in the armed services cannot omit the concept of 'cultural hybridization' in Dutch society in general and in the Dutch armed forces in particular. Employees of the future in the forces can no longer be one-dimensionally stereotyped as 'Dutch', 'Turkish', 'Moroccan', 'Antillean', 'woman' or 'man'. The entanglement of different cultures within individuals and groups of individuals demands another vision on the position of women and ethnic minorities in the armed services. Emancipation is good, but real diversity may be a more realistic future.

Notes

The authors wish to thank Dr. Twan Hendricks for his help in 'Churchillizing' this text.

1 Ministry of Defence, *Plan voor positieve actie voor integratie van de vrouw in de Krijgsmacht* (Plan for affirmative action for women in the armed forces), The Hague, 1989.
2 E. Ketting and K. Soesbeek (eds), *Homoseksualiteit en Krijgsmacht* (Homosexuality and armed forces), Delft: Eburon, 1992.
3 P.A. Stoppelenburg and G.J. Feenstra, *De positie van homoseksuelen bij Defensie. Een vervolgonderzoek naar de positie van homoseksuelen binnen de Defensieorganisatie* (The position of homosexuals in the defence organisation. Sequal research on the position of homosexuals in the defence organisation), Tilburg: IVA, 1999.
4 CBS, *Allochtonen in Nederland* (Ethnic minorities in the Netherlands), Voorburg, the Netherlands: CBS, 2004.
5 R. Richardson and J. Soeters, 'Managing Diversity in de Krijgsmacht', *Militaire Spectator*, 176, 3, 1998, 136–145; R. Richardson and J. Bosch, 'The Diversity Climate in the Dutch Armed Forces', in J. Soeters and J. van der Meulen (eds), *Managing Diversity in the Armed Forces*, Tilburg: Tilburg University Press, 1999, pp. 127–155.
6 CBS, *Allochtonen in Nederland*.
7 Ibid.
8 Ibid. An ethnic minority member in the Netherlands is defined as follows:

   1 A person born in Turkey, Morocco, Surinam, the Netherlands Antilles, Aruba, former Yugoslavia, or other countries in South or Central America, Africa or Asia with the exception of Japan and the former Dutch East Indies. Thus, people born after 27 December 1949 in Indonesia belong to the ethnic minority group.
   2 A person who can be found in the register, referred to as in article 1, of the *Wet Rietkerk* payment. (The Mollucan population group is included in this definition as well. Although they cannot be registered according to the principle of native country, they are included if they are entered in the register that has been used in the scope of the *Wet Rietkerk* payment. This register recorded people who were brought to the Netherlands in 1951 and 1952. Children of those registered are included in the definition as well.)
   3 Children of people denominated in 1 and 2.

9 Ibid.
10 Richardson and Bosch, 'The Diversity Climate'.
11 Of course, the murder of Pim Fortuyn on 6 May 2002 contributed to a tense atmosphere. While Fortuyn was a strong critic of Islam, the fact that his murderer was native Dutch gave this violent incident different overtones as compared with the murder of Van Gogh. However, Fortuyn's political heritage, his views on Islam and ethnic minorities very much included, are still making themselves felt in Dutch politics.
12 J. Latten, *Zwanger van segregatie* (Pregnant of segregation), Amsterdam: Vossiuspers, 2005.
13 Ministry of Social Affairs and Employment, *Handleiding voor de werkgever. Wet stimulering arbeidsdeelname minderheden* (Manual for the employer. Law on the activation of labour participation of ethnic minorities), The Hague: SZW, 1998.
14 Colourview, *Potentieel aan kleur? Een studie naar belangstelling van de allochtone jongeren voor een baan bij de krijgsmacht* (Potential in colour? Research in the interest of ethnic minorities in a job in the armed forces), The Hague: Ministry of Defence, 2000.
15 The core of this agreement between the Turkish and Dutch defence departments is

that Dutch-Turkish recruits, who also have to undergo eighteen months' conscription in the Turkish forces because of their double nationality, can buy off fifteen months of this conscription when they join the Dutch armed forces on a contract basis of three years. The Dutch defence department pays in advance, and the payments are deducted from the salary of the recruit on a monthly basis over these three years.
16 Koninklijke Luchtmacht, *Opzoeken*, The Hague: Ministry of Defence, 2005.
17 F. Bosman, R. Richardson, N. Guns and J. Ten Heuve, *Etnisch culturele diversiteit in de defensieorganisatie* (Ethnic cultural diversity in the defence organization), Breda: Netherlands Defence Academy, 2006.
18 C.C. Choenni, *Kleur in de Krijgsmacht. De integratie van jonge Surinaamse mannen in Nederland* (Colour in the Dutch armed forces: the integration of young male Surinamese in the Netherlands), Utrecht: ISOR, 1995.
19 F. Bosman, R. Richardson and J. Soeters, *Intercultural Tensions in the Military? Evidence from the Netherlands Armed Forces*, forthcoming.
20 J.W. Berry and R. Kalin, 'Multicultural and Ethnic Attitudes in Canada: An Overview of the 1991 National Survey', *Canadian Journal of Behavioural Science*, 27, 3, 1995, 301–320; J.W. Berry, Y.H. Poortinga, P. Segall and P. Daren, *Cross-cultural Psychology: Research and Applications*, Cambridge: Cambridge University Press, 1992.
21 The *assimilation* strategy is defined as 'when an acculturating individual does not wish to maintain culture and identity and seeks daily interaction with the dominant society'. The *separation* strategy is a strategy when 'there is a value placed holding onto one's original culture and wish to avoid interaction with others'. Berry et al., *Cross-cultural Psychology*, p. 278.
22 W. Portegijs, A. Boelens and L. Olsthoorn, *Emancipatiemonitor*, The Hague: Sociaal en Cultureel Planbureau/Centraal Bureau voor de Statistiek, 2004, also published on the internet: www.scp.nl/publicaties/boeken/9037701906.shtml.
23 Ibid.
24 This is officially acknowledged to be the date of entry of the first female soldiers into the Royal Netherlands armed forces. Actually, the Women's Corps of the Royal Netherlands Indian Army (the colonial army of the Netherlands stationed in the East Indies) was established earlier, on 5 March 1944. See S. Kruyswijk-van Thiel, *Het vrouwenkorps-Knil* (Women's Corps, Royal Netherlands Indian Army), dissertation based on qualitative interviews and quantitative research, Amsterdam: Dutch University Press, 2004, p. 12.
25 D. Iskra, S. Trainor, M. Leithauser and M.W. Segal, 'Women's Participation in Armed Forces Cross-nationally: Expanding Segal's Model', *Current Sociology*, 50, 5, 2002, 771–797.
26 H. Carreiras, *Gender and the Military: A Comparative Study of the Participation of Women in the Armed Forces of Western Democracies*, Florence: European University Institute, 2004, p. 229.
27 Bouta and Frerks distinguish seven roles of women: (1) Women as victims of (sexual) violence; (2) women as combatants; (3) women for peace in the non-governmental sector; (4) women in 'formal peace politics'; (5) women as coping and surviving actors; (6) women as household heads; and (7) women and (in)formal employment. Depending on the phase the conflict is in – pre-conflict, during the conflict, rebuilding society post-conflict – the possible interventions from women in these seven roles are different. Women may be guerrilla fighters, victims, survivors, nurses, NGO relief workers, military peacekeepers, UN observers, businesswomen starting up new enterprises, etc. See T. Bouta and G. Frerks, *Women's Multifaceted Role in Armed Conflict: Review of Selected Literature*, The Hague: Netherlands Institute of International Relations Clingendael, 2002.
28 In the spring of 2006 a former female sailor came forward with a shocking story about harassment and even rape on board a Navy frigate. A thorough investigation has been ordered by the State Secretary.

29 www.pvda.nl/renderer.do/menuId/37298/clearState/true/sf/37298/returnPage/37298/itemId/200035953/realItemId/200035953/pageId/45641/instanceId/37907 (accessed 16 November 2005).
30 J. Bosch and D.E.M. Verweij, 'Enduring Ambivalence: The Dutch Armed Forces and Its Women Recruits', in C. Cockburn and D. Zarkov (eds), *The Postwar Moment: Militaries, Masculinities and International Peacekeeping*, London: Lawrence & Wishart, 2002.
31 L. Sion, 'Changing from Green to Blue Beret: A Tale of Two Dutch Peacekeeping Units', Ph.D. thesis, Free University of Amsterdam, 2004.
32 The network can be found on the Internet at www.defensievrouwennetwerk.nl.
33 H. Carreiras, *Gender and the Military*, pp. 318–325.
34 J.D. Yoder, J. Adams and H. Prince, 'The Price of a Token', *Journal of Political and Military Sociology*, 11, 1983, 325–337.
35 F. Glastra, *Organisatie en Diversiteit. Naar een contextuele benadering van intercultureel management* (Organization and diversity. Towards a contextual approach of intercultural management), Utrecht: Lemma, 1999.
36 F. Bosman et al., *Etrisch culturele diversiteit*.
37 W. Welsch, 'Transculturality: The Puzzling Form of Cultures Today', in M. Featherstone and S. Lash (eds), *Spaces of Culture*, London: Sage, 1999, pp. 194–213.

Index

accommodation 49, 50, 54, 56, 58, 59, 80, 85, 155, 168
acculturation 48, 50, 126, 205; *see also* assimilation
affirmative action 19, 28, 77, 79, 80, 83, 84, 86, 90, 92, 92n5, 93n15, 132, 161, 180, 195, 212; *see also* equality
assimilation 46n53, 48, 113, 126, 155, 181, 202, 205, 213n21

citizenship 1, 2, 4, 6, 9, 13n19, 31, 34, 36, 48, 50–54, 56–58, 63n29, 68, 78, 102, 103, 107, 126–129, 131, 135–137, 138n9, 138n12, 140, 141, 144, 145, 147–149, 151, 155–157, 159–161, 166–168, 171–175, 182, 185, 187–189, 197n8, 197n12, 202
class 5, 11, 17, 52, 53, 64, 65, 71–73, 74, 76n20, 90, 96, 111, 114, 117, 119, 135, 145, 147, 155, 156, 193; class regiment in Indian army 144; social mobility 72
combat(ant) 13n23, 15, 21–23, 27, 30n12, 31–34, 37–44, 45n37, 46n46, 46n51, 46n53, 52, 57, 63n33, 80, 83, 86–88, 98–100, 105, 107, 108, 108n8, 109n10, 127, 132, 133, 156, 157, 162–164, 166, 170n9, 170n12, 179–181, 190, 191, 193, 198n17, 206, 208, 209, 213n27
conscription 2, 5, 8, 11, 16, 21, 50–59, 63, 67, 80, 101, 102, 104, 106–108, 127–129, 132, 134, 136, 138n7, 142, 154, 156, 158, 162, 163, 170n12, 175, 177, 180, 182, 200, 203, 207, 213; draft(ees) 16, 27, 51, 53, 101, 127, 132, 154, 159, 161, 182, 186, 187; *see also* professionalization; recruitment
constitution 33, 34, 53, 55–57, 65, 79, 80, 86, 87, 89, 92n9, 113, 116, 149, 156, 180, 188, 195, 197n9, 197n12; *see also* law

democracy 4, 12n14, 18, 41, 48, 49, 54, 58, 61, 66, 75n3, 77, 80–82, 87, 89, 111, 113, 115, 116, 122, 125, 126, 128, 138n18, 155, 161, 178, 209
demographics 2, 11, *35*, 44, 45n16, 64, 66, 76n10, 77, 81, 96, 107, 111, 113, 144, 145, 179, 201, 210
discrimination 32, 159, 160; class 146; ethnic 43, 55, 56, 60, 141, 154, 155, 167, 168; gender 60, 179, 190, 210; linguistic 187; policies 33–34, 58, 78–80, 86, 89, 140, 179, 190, 195; 'positive discrimination' 211; racial 16, 18, 19, 28, 68, 84; religious 90; sexual 87, 106, 162; 'zero tolerance' *27*, 28
draft *see* conscription

education 4, 5, 18, 20, 28, 30n7, 37, 55, 56, 58, 59, 72, 77, 78, 80, 83, 89–91, 93n17, 96, 98, 100–104, 106, 107, 110n33, 122, 129, 132, 138n18, 143, 145, 146, 152n23, 152n24, 152n25, 153n35, 164, 176, 191, 193, 201, 202, 205, 206, 211
effectiveness 2, 6, 8, 9, 20, 32, 39, 41, 46n46, 46n53, 48, 56, 58, 78, 80, 83, 88, 91, 92, 100, 101, 116, 117, 133, 140, 150, 159, 161, 163, 166, 169n7, 170n9, 182, 205, 210
emancipation: of ethnic minorities 211; of women 3, 104, 155, 158, 166, 178, 211
equality *see* ethnic equality; gender equality; racial equality; religion; sexual equality; sexual orientation
ethnic equality 96

family 25, 33, 52, 63n34, 64, 84, 96, 98, 103, 106, 117, 118, 120, 122, 145, 158, 164, 165, 182, 183, 196, 201, 208

216   Index

feminism 23, 30n12, 109n28, 128, 154, 155, 158, 160, 180

gays and lesbians *see* sexual orientation
gender equality 3; in Britain 147; in Canada 34; in Eritrea 96, 98–101, 104; in France 165–166; in Germany 179–182; in Israel 133–134; in the Netherlands 209; in South Africa 79, 85–87
geographical *see* region
glass ceiling 59, 133, 147, 206, 211

homosexual *see* sexual orientation
identity 4, 5, 11, 13n18, 13n27, 28, 29, 31, 49, 53, 55, 56, 61, 63n27, 68, 111, 116, 123, 123n2, 125–127, 135, 150, 161, 162, 164–166, 168, 169n7, 193, 197n9, 213n21
immigrant (immigration) 4, 17, 36, 68, 125, 126, 130–132, 137, 138n17, 154–156, 158, 160, 166, 168, 170n10, 172, 173, 201

language 11, 35, 37, 55, 56, 59, 78–80, 84, 85, 89, 91, 93n19, 96, 108n3, 111–113, 117, 118, 121, 126, 131, 172, 174–176, 182, 183, 185–187, 189, 193, 197, 197n14, 199n38, 199n39
law 2, 8, 25, 30n15, 31, 33, 34, 41, 42, 49, 51–53, 57, 71, 77–79, 87, 89, 90, 91, 98, 100, 103, 108, 109n13, 128, 133, 141–143, 149, 155, 162–164, 171, 172, 176–178, 180–183, 184n4, 186, 188, 190, 195, 197n12, 198n20, 203, 207, 212n13
leadership 4, 9, 12, 12n15, 13n32, 19, 20, 30n5, 33, 34, 40, 42, 43, 44n2, 45n16, 46n53, 50, 51, 53, 58–60, 63n32, 76n15, 82, 86, 87, 89, 91, 95, 97, 99–101, 104, 107, 115, 116, 126, 135, 142, 150, 164, 167, 169, 175, 176, 183, 209
legitimacy 5, 6, 41, 52, 80, 129, 132, 142, 143, 159, 163, 165

management 1, 2, 4, 7, 8, 11, 12n17, 13n30, 14n37, 14n40, 16, 39, 40, 52, 77–80, 83, 88–92, 92n11, 94n55, 94n57, 101, 110n65, 125, 129, 137, 154, 177, 185, 194, 199n37, 210, 211, 214n35
'martial' races, use of 114
media 17, 24, 25, 59, 63n32, 122, 137, 162, 168
military service 13n20, 34, 42, 44n8, 46n56, 51, 53, 57, 101, 102, 105–108, 126–128, 131, 132, 135–137, 138n4, 138n10, 138n21, 139n34, 147, 152n28, 174, 177, 178, 180, 182, 205, *207*
modernization 4, 9, 13n20, 27, 30n24, 37, 53, 62n19, 65, 75n3, 76n14, 98, 112, 113, 117, 118, 122, 125, 147, 154, 156, 159–163, 167–169, 185
multicultural(ism) 1, 2, 9, 13n19, 31, 43, 45n22, 49, 55–58, 60, 61, 63n29, 88, 116, 121, 125, 154, 155, 182, 196, 201, 202, 205, 206, 211, 213n20

occupation 29–32, 34, 38–41, 71, 75, 95, 98, 114, 128, 135, 140, 142, 145, 146–148, 150, 151, 164, 181, *192*, 193, *209*

participation 12n11, 20, 22, 31, 40, 45n16, 52, 59, 61, 65, 67, 75, 96, 97–100, 108n5, 109n22, 135, 136, 148, 152n28, 178, 179, 204, 206, 209, 211, 212n13, 213n25, 213n26
peacekeeping 13n24, 28, 44, 82, 87, 116, 123, 150, 185, 197, 208
politics 1, 3–5, 7, 9, 11, 12, 13n19, 48, 50, 51, 53, 54, 56, 57, 60, 61n3, 63n27, 65, *71*, 72, 74, 75n2, 75n3, 76n16, 76n17, 79, 82, 83, 86, 87, 92, 92n5, 93n31, 93n32, 96, 98, 99, 101, 104, 107, 108, 109n27, 114, 115, 117, 121, 122, 125, 132, 133, 135, 138n3, 138n6, 150, 155, 156, 159, 160, 161, 168, 169n7, 178–180, 183, 194, 202, 204, 212n11, 213n27, 214n34
professional(ization) 5, 33, 51, 52, 55, 63n30, 73, 76n17, 80, 82, 83, 89, 93n32, 107, 111, 112–116, 118, 122, 124n8, 124n12, 125, 128, 129, 131–137, 145, 154, 172, 176, 178, 190

quota 19, 35, 49, 83, 157, 173, 186

racial equality: in Britain 140–141; in South Africa 78, 79, 84
region 5, 11, 48, 49, 52–54, 59, 61, 68–70, 85, 93n36, 95, 96, 113, 117–119, 146, 170, 185, 186, 201; geographical 64, 69, 70, 113, 134, 145; rural 13n20, 17, 52, 55, 62, 95, 103; (sub)urban 52, 74, 111, 155, 167, 201
recruitment: Belgium 188–189; criteria 114; of ethnic minorities 50, 141–145, 147–148, 151n8; goals 54n14; Israel 127; Netherlands 203–204; policy 36,

Index   217

203; trends 44, 74; of women 120, 207–210, 163
religion 1, 5, 11, 33, 38, 59, 64, 65, 70, 71, 75, 79, 80, 84, 87, 89, 90, 94n33, 95–98, 101–103, 111–114, 116–119, 121, 122, 126–128, 130, 134–137, 139n35, 162, 167, 169n2, 170n9, 174–176, 182, 183, 184n6, 190, 202, 205
rural *see* region

segregation 9, 15, 16, 27, 69, 84, 92n5, 146, 155, 202, 212n12; separation 9, 13n35, 37, 122, 205
separation *see* segregation
sexual harassment 3, 16, 23, 24, 26, 27, 28, 30n9, 30n17, 32, 34, 46n53, 80, 90, 100, 102, 105, 106, 133, 141, 160, 161, 165, 167, 193–195, 209, 213n28; sexual abuse 23, 105, 106; 'zero tolerance' 28
sexual orientation 11, 14n39, 34, 46n60, 79, 80, 87, 155, 158, 177, 178, 185, 189–192; gays (and lesbians) 4, 11, 14n39, 31, 35, 41, 42, 67, 87, 88, 154–156, 158–160, 162, 163, 169, 177, 178, 101, 192, 200; homosexual 11, 31, 34, 41, 42, 67, 68, 75, 75n5, 85, 87, 88, 94n52, 94n53, 154, 155, 158, 162, 163, 168, 169, 169n6, 171, 177, 178, 183, 184n8, 185, 197n1, 200, 212n2, 212n3

socialization 73, 74, 76n17, 90, 104, 112, 117, 118, 121, 136, 137, 176, 210

training 18, 20–23, 25, 28, 30n9, 30n18, 39, 40, 43, 44, 45n20, 50, 86, 89, 90, 93n17, 94n54, 96, 101, 106, 108, 118, 120, 160, 166, 169n8, 181, 186, 190, 193–196, 204–206, 209–211

urban *see* region

youth 5, 10, 20, 58, 65, 76n15, 85, 95, 99, 103–106, 108, 111, 115, 120, 129, 136, 137, 142, 146, 157, 166, 168, 169, 182, 183, 187, 188, 202, 203, 210, 213n18

war 170n9; Bolivia and Ecuador 51–53, 56, 57; Cold War 142; conduct 1; guerrilla warfare 108n1; Eritrea 95, 97, 98, 100, 101, 102, 105–106, 107–108; First World War 8, 37, 115, 148, 186; Iraq War 8, 16–17, 23, 149, 152n15; Korean War 16, 27, 37; Second World War 2, 15, 21, 27, 37, 38, 40, 69, 115, 148; 'three-block war' 43–44; Vietnam War 16–17; warrior 12n10, 41, 46n45, 114, 181; *see also* peacekeeping
warrior *see* war

For Product Safety Concerns and Information please contact our EU representative GPSR@taylorandfrancis.com
Taylor & Francis Verlag GmbH, Kaufingerstraße 24, 80331 München, Germany